Student Solutions Manual

to accompany

Statistics:

Unlocking the Power of Data

Second Edition

Robin H. Lock

St. Lawrence University

Patti Frazer Lock

St. Lawrence University

Kari Lock Morgan

Pennsylvania State University

Eric F. Lock

University of Minnesota

Dennis F. Lock

Miami Dolphins

To order books or for customer service please, call 1-800-CALL WILEY (225-5945).

ISBN-13   978-1-119-30891-1

Printed in the United States of America

V10002009_062618

CONTENTS

CONTENTS

Section 1.1 Solutions

1.1 (a) The cases are the people who are asked the question.

(b) The variable is whether each person supports the law or not. It is categorical.

1.3 (a) The cases are the teenagers in the sample.

(b) The variable is the result (yes or no) indicating whether each teenager eats at least five servings a day of fruits and vegetables. It is categorical.

1.5 (a) The 10 beams that were tested.

(b) The force at which each beam broke. It is quantitative.

1.7 Since we expect the number of years smoking cigarettes to impact lung capacity, we think of the number of years smoking as the explanatory variable and the lung capacity as the response variable.

1.9 Ingesting more alcoholic drinks will cause the level of alcohol in the blood to increase, so the number of drinks is the explanatory variable and blood alcohol content is the response.

1.11 (a) *Year* and *HigherSAT* are categorical. The other six variables are all quantitative, although *Siblings* might be classified as either categorical or quantitative.

(b) There are many possible answers, such as "What proportion of the students are first year students?" or "What is the average weight of these students?"

(c) There are many possible answers, such as "Do seniors seem to weigh more than first year students?" or "Do students with high Verbal SAT scores seem to also have high Math SAT scores?"

1.13 There are at least two variables. One variable is whether or not the spider engaged in mock-sex. This variable is categorical and the explanatory variable. Another variable is length of time to reach the point of real mating once the spider is fully mature. This variable is quantitative and the response variable.

1.15 There are two variables. One variable indicates the presence or absence of the gene variant and the second variable indicates which of the three ethnic groups the individual belongs to. Both variables are categorical.

1.17 (a) There are 10 cases, corresponding to the 10 cities. The two variables are population, which is quantitative, and the hemisphere the city is in, which is categorical.

(b) We need two columns, one for each variable. The columns can be in either order. See the table.

Population	Hemisphere
37	Eastern
26	Eastern
23	Eastern
22	Eastern
21	Eastern
21	Eastern
21	Eastern
21	Western
20	Western
19	Western

1.19 One variable is whether the young female mice lived in an enriched environment or not. This is the explanatory variable and it is categorical. The response variable is how fast the offspring learned to navigate mazes and is quantitative.

1.21 (a) This description of the study mentions six variables: age, nose volume, nose surface area, nose height, nose width, and gender.

(b) One of the variables (gender) is categorical, and the other five are quantitative.

(c) There are six variables so the dataset will have six columns. The 859 participants are the cases, so the dataset will have 859 rows.

1.23 (a) The cases are the 40 people with insomnia who were included in the study.

(b) There are two variables. One is which group the person is assigned to, either therapy or not, and the other is whether or not the person reported sleep improvements. Both are categorical.

(c) The dataset would have two columns, one for each of the two variables, and 40 rows, one for each of the people in the study.

1.25 We could sample people eligible to vote and ask them each their political party and whether they voted in the last election. The cases would be people eligible to vote that we collect data from. The variables would be political party and whether or not the person voted in the last election. Alternatively, we could ask whether each person plans to vote in an upcoming election.

1.27 Answers will vary.

Section 1.2 Solutions

1.29 This is a population, because all customers are accounted for.

1.31 This is a sample, because only a subset of college students were sent the questionnaire.

1.33 The sample is the five hundred Canadian adults that were asked the question; the population is all Canadian adults.

1.35 The sample is the 1000 households which have databoxes attached to the televisions. The population is all US households with televisions.

1.37   (a) The sample is the 10 selected twitter accounts.

  (b) The target population is all twitter accounts.

  (c) The population we can generalize to, given the sample, is only twitter accounts of this author's followers, since this is the population from which the sample was drawn.

1.39   (a) The sample is the girls who are on the selected basketball teams.

  (b) The population we are interested in is all female high school students.

  (c) A population we can generalize to, given our sample, is female high school students who are on a basketball team.

1.41 Yes, this is random sample from the population.

1.43 No, this is not a random sample, because certain segments of the population (e.g. those not attending college) cannot be selected.

1.45 No, this is not a random sample, this is a volunteer sample, since the only people in the sample are those that self-select to respond to the online poll.

1.47 This is biased because the way the question is worded is not at all objective. Although the sample is a random sample, the wording bias may distort the results.

1.49 From the description, it appears that this method of data collection is not biased.

1.51 Because this was a random sample of parents in Kansas City, the result can be generalized to all parents in Kansas City.

1.53   (a) Yes, the sample is likely to be representative since it is a random sample.

  (b) Yes, since the sample is a random sample, we can generalize to the population of all Canadian consumers.

1.55   (a) The individual cases are the over 6000 restroom patrons who were observed. The description makes it clear that at least three variables are recorded. One is whether or not the person washed their hands, another is the gender of the individual, and a third is the location of the observation. All three are categorical.

  (b) In a phone survey, people are likely to represent themselves in the best light and not always give completely honest answers. That is why it is important to also find other ways of collecting data, such as this method of observing people's actual habits in the restroom.

1.57 No. This is a volunteer sample, and there is reason to believe the participants are not representative of the population. For example, some may choose to participate because they LIKE alcohol and/or marijuana, and those in the sample may tend to have more experience with these substances than the overall population. In addition, the advertisements for the study were aired on rock radio stations in Sydney, so only those people who listen to rock radio stations in Sydney would hear about the option to participate.

1.59 (a) This is not a simple random sample from the population, since only those who saw and wanted to click and complete the survey were included.

(b) These results could also have been biased by how the survey was constructed. The wording of the questions might also introduce bias.

1.61 The sample of planes that return from bombing missions was biased. More bullet holes were found in the wings and tail because planes that were shot in other regions were more likely to crash and not return.

1.63 (a) Since the NHANES sample is drawn from all people in the US, that is the population we can generalize to.

(b) Since the NHAMCS sample is drawn from patients in emergency rooms in the US, we can generalize the results to all emergency room patients in the US.

(c) i. NHANES: The question about an association between being overweight and developing diabetes applies to all people in the US, not just those who visit an emergency room.

ii. NHAMCS: This question asks specifically about the type of injury for people who go to an emergency room.

iii. NHAMCS: This question of average waiting time only applies to emergency room patients.

iv. NHANES: This question is asking about all US residents. Note that the proportion would be equal to one for the people sampled in NAMCS since they only get into the sample if they visit an emergency room!

1.65 Answers will vary. See the technology notes to see how to use specific technology to select a random sample.

Section 1.3 Solutions

1.67 Since "no link is found" there is neither association nor causation.

1.69 The phrase "more likely" indicates an association, but there is no claim that wealth *causes* people to lie, cheat or steal.

1.71 The statements imply that eating more fiber will cause people to lose wait, so this is a causal association.

1.73 One possible confounding variable is population. Increasing population in the world over time may mean more beef and more pork is consumed. Other answers are possible. Remember that a confounding variable should be associated with both of the variables of interest.

1.75 One possible confounding variable is snow in the winter. When there is more snow, sales of both toboggans and mittens will be higher. Remember that a confounding variable should be associated with both of the variables of interest.

1.77 One possible confounding variable is gender. Males usually have shorter hair and are taller. Other answers are possible. Remember that a confounding variable should be associated with both of the variables of interest.

1.79 We are actively manipulating the explanatory variable (playing music or not), so this is an experiment.

1.81 We are not manipulating any variables in this study, we are only measuring things (omega-3 oils and water acidity) as they exist. This is an observational study.

1.83 The penguins in this study were randomly assigned to get either a metal or an electronic tag so this is an experiment.

1.85 A possible confounding variable is amount of snow and ice on the roads. When more snow and ice has fallen, more salt will be needed *and* more people will have accidents. Notice that the confounding variable has an association with *both* the variables of interest.

1.87 Yes, this study provides evidence that louder music causes people to drink more beer, because the explanatory variable (volume of music) was randomly determined by the researchers and an association was found.

1.89 (a) Yes. Because the study in mice was a randomized experiment, we can conclude causation.

   (b) No. Since it appears that the study in humans was an observational study, it is not appropriate to conclude causation. Although the headline may still be true for humans, we cannot make this conclusion based on the study described.

1.91 (a) The cases are the 2,623 schoolchildren.

   (b) The explanatory variable is the amount of greenery around the schools.

   (c) The response variable is the score on the memory and attention tests.

   (d) Yes, the headline implies that more green space causes kids to be smarter.

   (e) No variables were manipulated so this is an observational study.

   (f) No! Since this is not an experiment, we cannot conclude causation.

(g) The socioeconomic status of the children is a possible confounding variable, since it is likely to effect both the amount of green space and also the test scores. There are other possible answers.

1.93 (a) The explanatoy variable is amount of leisure time spent sitting and the response variable is whether or not the person gets cancer.

(b) This is an observational study because the explanatory variable was not randomly assigned.

(c) No, we cannot conclude spending more leisure time sitting causes cancer in women because this is an observational study.

(d) No, we also cannot conclude that spending more leisure time sitting does not cause cancer in women; because this was an observational study we can make no conclusions about causality. Sitting may or may not cause cancer.

1.95 (a) This is an observational study because the explanatory variable (time spent on affection after sex) was not randomly assigned.

(b) No, because this is an observational study. It's quite possible that people in stronger, more loving relationships simply tend to spend more time cuddling after sex, not that cuddling after sex causes relationship happiness.

(c) No, the phrase "boosts" implies a causal relationship, which cannot be supported by an observational study.

(d) No, the phrase "promotes" implies a causal relationship, which cannot be supported by an observational study.

1.97 (a) The explanatory variable is whether the person just had a full night of sleep or 24 hours of being awake. The response variable is ability to recognize facial expressions.

(b) This is a randomized experiment, a matched pairs experiment because each person received both treatments.

(c) Yes, we can conclude that missing a night of sleep hinders the ability to recognize facial expressions, because the explanatory variable was randomly assigned.

(d) No, we cannot conclude that better quality of REM sleep improves ability to recognize facial expressions, because the explanatory variable in this case (quality of REM sleep) was not randomly assigned.

1.99 (a) This is an experiment since the background color was actively assigned by the researchers.

(b) The explanatory variable is the background color, which is categorical. The response variable is the attractiveness rating, which is quantitative.

(c) The men were randomly divided into the two groups. Blinding was used by not telling the participants or those working with them the purpose of the study.

(d) Yes. Since this was a well-designed randomized experiment, we can conclude that there is a causal relationship.

1.101 The explanatory variables are the type of *payment* and *sex*. Only the type of payment can be randomly assigned. The number of *items* ordered and *cost* are response variables.

1.103 (a) We randomly divide the participants into two groups of 25 each. Half will be given fluoxetine and half will get a placebo.

(b) The placebo pills will look exactly like the fluoxetine pills and will be taken the same way, but they will not have any active ingredients.

(c) The patients won't know who is getting which type of pill (the fluoxetine or the placebo) and the people treating the patients and administering the questionnaire won't know who is in which group.

1.105 (a) Randomly assign 25 people to carbo-load and 25 people to not carbo-load and then measure each person's athletic performance the following day.

(b) We would have each person carbo-load and not carbo-load, on different days (preferably different weeks). The order would be randomly determined, so some people would carbo-load first and other people would carbo-load second. In both cases athletic performance would be measured the following day and we would look at the difference in performance for each person between the two treatments.

(c) The matched pairs experiment is probably better because we are able to compare the different effects for the same person. It is more precise comparing one person's athletic performance under two different treatments, rather than different people's athletic performance under two different treatments.

1.107 Answers will vary. Example: The total amount of pizza consumed and the total amount of cheese consumed, per year, over the last century. Eating more pizza causes people to eat more cheese, but the overall rise in population is also a confounding variable.

Section 2.1 Solutions

2.1 The total number is $169 + 193 = 362$, so we have $\hat{p} = 169/362 = 0.4669$. We see that 46.69% are female.

2.3 The total number is $94 + 195 + 35 + 36 = 360$ and the number who are juniors or seniors is $35 + 36 = 71$. We have $\hat{p} = 71/360 = 0.1972$. We see that 19.72% percent of the students who identified their class year are juniors or seniors.

2.5 Since this describes a proportion for all residents of the US, the proportion is for a population and the correct notation is p. We see that the proportion of US residents who are foreign born is $p = 0.124$.

2.7 The report describes the results of a sample, so the correct notation is $\hat{p}$. The proportion of US teens who say they have made a new friend online is $\hat{p} = 605/1060 = 0.571$.

2.9 A relative frequency table is a table showing the proportion in each category. We see that the proportion preferring an Academy award is $31/362 = 0.086$, the proportion preferring a Nobel prize is $149/362 = 0.412$, and the proportion preferring an Olympic gold medal is $182/362 = 0.503$. These are summarized in the relative frequency table below. In this case, the relative frequencies actually add to 1.001 due to round-off error.

Response	Relative Frequency
Academy award	0.086
Nobel prize	0.412
Olympic gold medal	0.503
Total	1.00

2.11 (a) We see that there are 200 cases total and 80 had Outcome A, so the proportion with Outcome A is $80/200 = 0.40$.

(b) We see that there are 200 cases total and 100 of them are in Group 1, so the proportion in Group 1 is $100/200 = 0.5$.

(c) There are 100 cases in Group 1, and 80 of these had Outcome B, so the proportion is $80/100 = 0.80$.

(d) We see that 80 of the cases had Outcome A and 60 of these were in Group 2, so the proportion is $60/80 = 0.75$.

2.13 Since the dataset includes all professional soccer games, this is a population. The cases are soccer games and there are approximately 66,000 of them. The variable is whether or not the home team won the game; it is categorical. The relevant statistic is $p = 0.624$.

2.15 (a) The sample is the 119 players who were observed. The population is all people who play rock-paper-scissors. The variable records which of the three options each player plays. This is a categorical variable.

(b) A relative frequency table is shown below. We see that rock is selected much more frequently than the others, and then paper, with scissors selected least often.

Option selected	Relative frequency
Rock	0.555
Paper	0.328
Scissors	0.118
Total	1.0

(c) Since rock is selected most often, your best bet is to play paper.

(d) Your opponent is likely to play paper again, so you should play scissors.

2.17 (a) The table is given.

	HS or less	Some college	College grad	Total
Agree	363	176	196	735
Disagree	557	466	789	1812
Don't know	20	26	32	78
Total	940	668	1017	2625

(b) For the survey participants with a high school degree or less, we see that $363/940 = 0.386$ or 38.6% agree. For those with some college, the proportion is $176/668 = 0.263$, or 26.3% agree, and for those with a college degree, the proportion is $196/1017 = 0.193$, or 19.3% agree. There appears to be an association, and it seems that as education level goes up, the proportion who agree that every person has one true love goes down.

(c) We see that $1017/2625 = 0.387$, or 38.7% of the survey responders have a college degree or higher.

(d) A total of 1812 people disagreed and 557 of those have a high school degree or less, so we have $557/1812 = 0.307$, or 30.7% of the people who disagree have a high school degree or less.

2.19 (a) The proportion of children who were given antibiotics is $438/616 = 0.711$.

(b) The proportion of children who were classified as overweight at age 9 is $181/616 = 0.294$.

(c) The proportion of those receiving antibiotics who were classified as overweight at age 9 is $144/438 = 0.329$.

(d) The proportion of those not receiving antibiotics who were classified as overweight at age 9 is $37/178 = 0.208$.

(e) Since $\hat{p}_A = 0.329$ and $\hat{p}_N = 0.208$, the difference in proportions is $\hat{p}_A - \hat{p}_N = 0.329 - 0.208 = 0.121$.

(f) Out of all children classified as overweight, the proportion who were given antibiotics is $144/181 = 0.796$.

2.21 Since these are population proportions, we use the notation p. We use p_H to represent the proportion of high school graduates unemployed and p_C to represent the proportion of college graduates (with a bachelor's degree) unemployed. (You might choose to use different subscripts, which is fine.) The difference in proportions is $p_H - p_C = 0.097 - 0.052 = 0.045$.

2.23 (a) This is an observational study since the researchers are observing the results after the fact and are not manipulating the gene directly to force a disruption. There are two variables: whether or not the person has dyslexia and whether or not the person has the DYXC1 break.

(b) Since $109 + 195 = 304$ people participated in the study, there will be 304 rows. Since there are two variables, there will be 2 columns: one for dyslexia or not and one for gene break or not.

(c) A two-way table showing the two groups and gene status is shown.

	Gene break	No break	Total
Dyslexia group	10	99	109
Control group	5	190	195
Total	15	289	304

(d) We look at each row (Dyslexia and Control) individually. For the dyslexia group, the proportion with the gene break is $10/109 = 0.092$. For the control group, the proportion with the gene break is $5/195 = 0.026$.

(e) There is a very substantial difference between the two proportions in part (d), so there appears to be an association between this particular genetic marker and dyslexia for the people in this sample. (As mentioned, we see in Chapter 4 how to determine whether we can generalize this result to the entire population.)

(f) We cannot assume a cause-and-effect relationship because this data comes from an observational study, not an experiment. There may be many confounding variables.

2.25   (a) This is an experiment. Participants were actively assigned to receive either electrical stimulation or sham stimulation.

(b) The study appears to be single-blind, since it explicitly states that participants did not know which group they were in. It is not clear from the description whether the study was double-blind.

(c) There are two variables. One is whether or not the participants solved the problem and the other is which treatment (electrical stimulation or sham stimulation) the participants received. Both are categorical.

(d) Since the groups are equally split, there are 20 participants in each group. We know that 20% of the control group solved the problem, and 20% of 20 is $0.20(20) = 4$ so 4 solved the problem and 16 did not. Similarly, in the electrical stimulation group, $0.6(20) = 12$ solved the problem and 8 did not. See the table.

Treatment	Solved	Not solved
Sham	4	16
Electrical	12	8

(e) We see that $4 + 12 = 16$ people correctly solved the problem, and 12 of the 16 were in the electrical stimulation group, so the answer is $12/16 = 0.75$. We see that 75% of the people who correctly solved the problem had the electrical stimulation.

(f) We have $\hat{p}_E = 0.60$ and $\hat{p}_S = 0.20$ so the difference in proportions is $\hat{p}_E - \hat{p}_S = 0.60 - 0.20 = 0.40$.

(g) The proportions who correctly solved the problem are quite different between the two groups, so electrical stimulation does seem to help people gain insight on a new problem type.

2.27   (a) The total number of respondents is 27,268 and the number answering zero is 18,712, so the proportion is $18712/27268 = 0.686$. We see that about 68.6% of respondents have not had five or more drinks in a single sitting at any time during the last two weeks.

(b) We see that 853 students answer five or more times and 495 of these are male, so the proportion is $495/853 = 0.580$. About 58% of those reporting that they drank five or more alcoholic drinks at least five times in the last two weeks are male.

(c) There are 8,956 males in the survey and $912 + 495 = 1407$ of them report that they have had five or more alcoholic drinks at least three times, so the proportion is $1407/8956 = 0.157$. About 15.7% of male college students report having five or more alcoholic drinks at least three times in the last two weeks.

(d) There are 18,312 females in the survey and $966 + 358 = 1324$ of them report that they have had five or more alcoholic drinks at least three times, so the proportion is $1324/18312 = 0.072$. About 7.2% of female college students report having five or more alcoholic drinks at least three times in the last two weeks.

2.29 (a) More females answered the survey since we see in graph (a) that the bar is much taller for females.

(b) It appears to be close to equal numbers saying they had no stress, since the height of the brown bars in graph (a) are similar. Graph (a) is the appropriate graph here since we are being asked about actual numbers not proportions.

(c) In this case, we are being asked about percents, so we use the relative frequencies in graph (b). We see in graph (b) that a greater percent of males said they had no stress.

(d) We are being asked about percents, so we use the relative frequencies in graph (b). We see in graph (b) that a greater percent of females said that stress had negatively affected their grades.

2.31 A two-way table to compare the participation rate of the *Reward* and *Deposit* groups is shown below.

Group	Accepted	Declined	Total
Reward	914	103	1017
Deposit	146	907	1053
Total	1060	1010	2070

To compare the participation rates between the two treatments, we find the proportion in each group who agreed to participate.

$$\text{Reward: } 914/1017 = 0.899 \text{ vs Deposit: } 146/1053 = 0.139$$

Not surprisingly, we see that the percentage in the *Reward* group who accepted the offer to participate in the program (89.9%) is much higher than in the *Deposit* group (13.9%) who were asked to risk some of their own money.

2.33 (a) If there is a clear association, then there is an obvious difference in the outcomes based on which treatment is used. There are many possible answers, but the most extreme difference (in which A is always successful and B never is) is shown below.

	Successful	Not successful	Total
Treatment A	20	0	20
Treatment B	0	20	20
Total	20	20	40

(b) If there is no association, then there is no difference in the outcomes between Treatments A and B. There are many possible answers, but in every case the Treatment A and Treatment B rows would be the same or very similar. One possibility is shown in table below.

	Successful	Not successful	Total
Treatment A	15	5	20
Treatment B	15	5	20
Total	30	10	40

2.35 (a) The *Year* variable in **StudentSurvey** has two missing values. Tallying the 360 nonmissing values gives the table below.

FirstYear	Sophomore	Junior	Senior
94	195	35	36

(b) The largest count is 195 sophomores. The relative frequency is $195/360 = 0.542$ or 54.2%.

2.37 Here is a two-way table showing the distribution of *Year* and *Gender*, with column percentages to show the gender breakdown in each class year.

	FirstYear	Junior	Senior	Sophomore	All
F	43	18	10	96	167
	45.74	51.43	27.78	49.23	46.39
M	51	17	26	99	193
	54.26	48.57	72.22	50.77	53.61
All	94	35	36	195	360
	100.00	100.00	100.00	100.00	100.00

We see that the male/female split is close to 50/50 for most of the years, except for the senior year which appears to have a much higher proportion of males.

2.39 Here is a side-by-side bar chart showing the relationship between class year and gender for the **StudentSurvey** data. You might also choose a stacked bar chart or ribbon plot to show the relationship. Note that the categories are ordered alphabetically, rather than in year sequence.

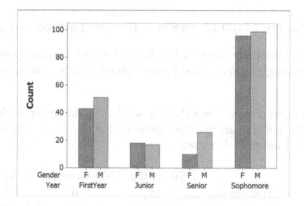

2.41 Graph (b) is the impostor. It shows more parochial students than private school students. The other three graphs have more private school students than parochial.

Section 2.2 Solutions

2.43 Only histogram F is skewed to the right.

2.45 While all of B,C,D,E and G are approximately symmetric, only B,C and E are also bell shaped.

2.47 Histograms E and G are both approximately symmetric, so the mean and median will be approximately equal. Histogram F is skewed right, so the mean should be larger then the median; while histogram H is skewed left, so the mean should be smaller then the median.

2.49 There are many possible dotplots we could draw that would be clearly skewed to the left. One is shown.

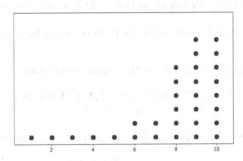

2.51 There are many possible dotplots we could draw that are approximately symmetric but not bell-shaped. One is shown.

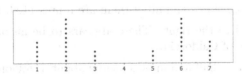

2.53 (a) We have $\bar{x} = (8 + 12 + 3 + 18 + 15)/5 = 11.2$.

(b) The median is the middle number when the numbers are put in order smallest to largest. In order, we have:

$$3 \quad 8 \quad 12 \quad 15 \quad 18.$$

The median is $m = 12$. Notice that there are two data values less than the median and two data values greater.

(c) There do not appear to be any particularly large or small data values relative to the rest, so there do not appear to be any outliers.

2.55   (a) We have $\bar{x} = (15 + 22 + 12 + 28 + 58 + 18 + 25 + 18)/8 = 24.5$.

(b) Since there are $n = 8$ values, the median is the average of the two middle numbers when the numbers are put in order smallest to largest. In order, we have:

$$12 \quad 15 \quad 18 \quad 18 \quad 22 \quad 25 \quad 28 \quad 58.$$

The median is the average of 18 and 22, so $m = 20$. Notice that there are four data values less than the median and four data values greater.

(c) The value 58 is significantly larger than all the other data values, so 58 is a likely outlier.

2.57   This is a sample, so the correct notation is $\bar{x} = 2386$ calories per day.

2.59   This is a population, so the correct notation is $\mu = 41.5$ yards per punt.

2.61   (a) We expect the mean to be larger since there appears to be a relatively large outlier (26.0) in the data values.

(b) There are eight numbers in the data set, so the mean is the sum of the values divided by 8. We have:

$$\text{Mean} = \frac{0.8 + 1.9 + 2.7 + 3.4 + 3.9 + 7.1 + 11.9 + 26.0}{8} = \frac{57.7}{8} = 7.2 \text{ mg/kg.}$$

The data values are already in order smallest to largest, and the median is the average of the two middle numbers. We have:

$$\text{Median} = \frac{3.4 + 3.9}{2} = 3.65.$$

2.63   (a) This is a mean. Since number of cats owned is always a whole number, a median of 2.39 is impossible.

(b) Since this is a right-skewed distribution, we expect the mean to be greater than the median.

2.65   (a) Since there are only 50 states and all of them are represented, this is the entire population.

(b) The distribution is skewed to the right. There appears to be an outlier at about 40 million. (The outlier represents the state of California.)

(c) The median splits the data in half and appears to be about 4 million. (In fact, it is 4.53 million.)

(d) The mean is the balance point for the histogram and is harder to estimate. It appears to be about 6 million. (In fact, it is 6.36 million.)

2.67   (a) The distribution is skewed to the left since there are many values between about 74 and 83 and then a long tail going down to the outliers on the left.

(b) Since half the values are above 74, the median is about 74. (The actual median is 73.8.)

(c) Since the data is skewed to the left, the mean will be less than the median so the mean will be less than 74. (The actual mean is 70.842.)

2.69   (a) The mean number of minutes on the treadmill for the mice receiving young blood is $\bar{x}_Y = 56.76$ minutes.

(b) The mean number of minutes on the treadmill for the mice receiving old blood is $\bar{x}_O = 34.69$ minutes.

(c) We see that $\bar{x}_Y - \bar{x}_O = 56.76 - 34.69 = 22.07$. The mice receiving young blood were able to run on the treadmill for 22 minutes longer, on average, than the mice receiving old blood.

(d) This is a randomized comparative experiment, as the mice were randomly assigned to the two groups.

(e) Yes, we can conclude causation since the data come from an experiment.

2.71 The notation for a median is m. We use m_H to represent the median earnings for high school graduates and m_C to represent the median earnings for college graduates. (You might choose to use different subscripts, which is fine.) The difference in medians is $m_H - m_C = 626 - 1025 = -399$. College graduates earn about $400 more per week than high school graduates.

2.73 (a) There are many possible answers. One way to force the outcome is to have a very small outlier, such as
2, 51, 52, 53, 54.
The median of these 5 numbers is 52 while the mean is 42.4.

(b) There are many possible answers. One way to force the outcome is to have a very large outlier, such as
2, 3, 4, 5, 200.
The median of these 5 numbers is 4 while the mean is 42.8.

(c) There are many possible answers. One option is the following:
2, 3, 4, 5, 6.
Both the mean and the median are 4.

2.75 The histogram is shown below.

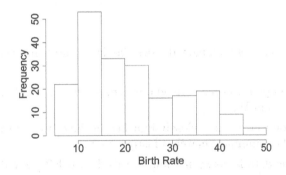

We see that it is strongly skewed to the right.

2.77 (a) It appears that the mean of the married women is higher than the mean of the never married women. We expect that the mean and the median will be the most different for the never married women, since that data is quite skewed while the married data is more symmetric.

(b) We have $n = 1000$ in each case. For the married women, we see that 162 women had 0 children, 213 had 1 child, and 344 had 2 children, so $162 + 213 + 344 = 719$ had 0, 1, or 2 children. Less than half the women had 0 or 1 child and more than half the women had 0, 1, or 2 children so the median is 2. For the never married women, more than half the women had 0 children, so the median is 0.

Section 2.3 Solutions

2.79 (a) Using technology, we see that the mean is $\overline{x} = 15.09$ with a standard deviation of $s = 13.30$.

(b) Using technology, we see that the five number summary is $(1, 4, 10, 25, 42)$. Notice that these five numbers divide the data into fourths.

2.81 (a) Using technology, we see that the mean is $\overline{x} = 59.73$ with a standard deviation of $s = 17.89$.

(b) Using technology, we see that the five number summary is $(25, 43, 64, 75, 80)$. Notice that these five numbers divide the data into fourths.

2.83 (a) Using technology, we see that the mean is $\overline{x} = 6.50$ hour per week with a standard deviation of $s = 5.58$.

(b) Using technology, we see that the five number summary is $(0, 3, 5, 9.5, 40)$. Notice that these five numbers divide the data into fourths.

2.85 Remember that a standard deviation is an approximate measure of the average distance of the data from the mean. Be sure to pay close attention to the scale on the horizontal axis for each histogram.

(a) V

(b) III

(c) IV

(d) I

(e) VI

(f) II

2.87 Remember that the five number summary divides the data (and hence the area in the histogram) into fourths.

(a) This shows a distribution pretty evenly spread out across the numbers 1 through 9, so this five number summary matches histogram W.

(b) This shows a distribution that is more closely bunched in the center, since 50% of the data is between 4 and 6. This five number summary matches histogram X.

(c) Since the top 50% of the data is between 7 and 9, this data is left skewed and matches histogram Y.

(d) Since both the minimum and the first quartile are 1, there is at least 25% of the data at 1, so this five number summary matches histogram Z.

2.89 The 10th-percentile is the value with 10% of the data values below it, so a reasonable estimate would be between 460 and 470. The 90th-percentile is the value with about 10% of the values above it, so a reasonable estimate would be between 530 and 540.

2.91 The mean appears to be about 68. Since the data is relatively bell-shaped, we can estimate the standard deviation using the 95% rule. Since there are 100 dots in the dotplot, we want to find the boundaries with 2 or 3 dots more extreme on either side. This gives boundaries from 59 to 76, which is 8 or 9 units above and below the mean. We estimate the standard deviation to be about 4.5.

2.93 We see that the minimum value is 58 and the maximum is 77. We can count the dots to find the value at the 25th-percentile, the 50th-percentile, and the 75th-percentile to find the quartiles and the median. We see that $Q_1 = 65$, the median is at 68, and $Q_3 = 70$. The five number summary is $(58, 65, 68, 70, 77)$.

2.95 For this dataset, half of the values are clustered between 100 and 115, and the other half are very spread out to the right between 115 and 220. This distribution is skewed to the right.

2.97 This data appears to be quite symmetric about the median of 36.3.

2.99 We have

$$Z\text{-score} = \frac{\text{Value} - \text{Mean}}{\text{Standard deviation}} = \frac{88 - 96}{10} = -0.8.$$

This value is 0.80 standard deviations below the mean, which is likely to be relatively near the center of the distribution.

2.101 We have

$$Z\text{-score} = \frac{\text{Value} - \text{Mean}}{\text{Standard deviation}} = \frac{8.1 - 5}{2} = 1.55.$$

This value is 1.55 standard deviations above the mean.

2.103 The 95% rule says that 95% of the data should be within two standard deviations of the mean, so the interval is:

$$\begin{array}{ccc}
\text{Mean} & \pm & 2 \cdot \text{StDev} \\
10 & \pm & 2 \cdot (3) \\
10 & \pm & 6 \\
4 & \text{to} & 16.
\end{array}$$

We expect 95% of the data to be between 4 and 16.

2.105 The 95% rule says that 95% of the data should be within two standard deviations of the mean, so the interval is:

$$\begin{array}{ccc}
\text{Mean} & \pm & 2 \cdot \text{StDev} \\
1500 & \pm & 2 \cdot (300) \\
1500 & \pm & 600 \\
900 & \text{to} & 2100.
\end{array}$$

We expect 95% of the data to be between 900 and 2100.

2.107 (a) We see in the computer output that the mean obesity rate is $\mu = 28.766\%$ and the standard deviation is $\sigma = 3.369\%$.

(b) We see that the largest value is 35.100, so we compute the z-score as:

$$z\text{-score} = \frac{x - \mu}{\sigma} = \frac{35.100 - 28.766}{3.369} = 1.880.$$

The maximum of 35.1% obese (which occurs for both Mississippi and West Virginia) is slightly less than two standard deviations above the mean.
We compute the z-score for the smallest percent obese, 21.3%, similarly:

$$z\text{-score} = \frac{x - \mu}{\sigma} = \frac{21.3 - 28.766}{3.369} = -2.216.$$

The minimum of 21.3% obese, from the state of Colorado, is about 2.2 standard deviations below the mean. The minimum might be considered a mild outlier.

(c) Since the distribution is relatively symmetric and bell-shaped, we expect that about 95% of the data will lie within two standard deviations of the mean. We have:

$$\mu - 2\sigma = 28.766 - 2(3.369) = 22.028 \quad \text{and} \quad \mu + 2\sigma = 28.766 + 2(3.369) = 35.504.$$

We expect about 95% of the data to lie between 22.028% and 35.504%. In fact, this general rule is very accurate in this case, since the percent of the population that is obese lies within this range for 48 of the 50 states, which is 96%. (The only states outside the range are Colorado and Hawaii, both outside the interval on the low side.)

2.109   (a) The z-score for the US will be positive because the value for the US is higher than the mean.

(b) $z = (5.2 - 4.7)/2 = 0.25$

(c) The range is $max - min = 13 - 0.8 = 12.2$.

(d) The IQR is $Q3 - Q1 = 5.6 - 3.2 = 2.4$.

2.111   (a) See the table.

Year	Joey	Takeru	Difference
2009	68	64	4
2008	59	59	0
2007	66	63	3
2006	52	54	-2
2005	32	49	-17

(b) For the five differences, we use technology to see that the mean is -2.4 and the standard deviation is 8.5.

2.113   (a) The 10th percentile is the value with 10% of the area of the histogram to the left of it. This appears to be at about 2.5 or 2.6. A (self-reported) grade point average of about 2.6 has 10% of the reported values below it (and 90% above). The 75th percentile appears to be at about 3.4. A grade point average of about 3.4 is greater than 75% of reported grade point averages.

(b) It appears that the highest GPA in the dataset is 4.0 and the lowest is 2.0, so the range is $4.0 - 2.0 = 2.0$.

2.115 Using technology, we see that the mean is 296.44 billion dollars and the standard deviation is 37.97 billion dollars. The 95% rule says that 95% of the data should be within two standard deviations of the mean, so the interval is:

$$
\begin{array}{ccc}
\text{Mean} & \pm & 2 \cdot \text{StDev} \\
296.44 & \pm & 2 \cdot (37.97) \\
296.44 & \pm & 75.94 \\
220.50 & \text{to} & 372.38.
\end{array}
$$

We expect 95% of US monthly retail sales over this time period to be between 220.50 billion dollars and 372.38 billion dollars.

2.117 (a) Using technology, we see that the mean is 38.6 blocks in a season and the standard deviation is 39.82 blocks.

(b) Using technology, we see that the five number summary is $(0, 14, 23, 48, 200)$.

(c) The five number summary from part (b) is more resistant to outliers and is often more appropriate if the data is heavily skewed.

(d) We create either a histogram, dotplot, or boxplot. A histogram of the data in *Blocks* is shown. We see that the distribution is heavily skewed to the right.

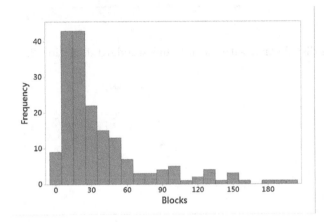

(e) This distribution is not at all bell-shaped, so it is not appropriate to use the 95% rule with this distribution.

2.119 (a) We calculate z-scores using the summary statistics for each: Critical Reading $= \frac{600-497}{115} = 0.896$, Math $= \frac{600-513}{120} = 0.725$, Writing $= \frac{600-487}{115} = 0.983$.

(b) Stanley's most unusual score was in the Writing component, since he has the highest z-score in this section. His least unusual score was in Mathematics.

(c) Stanley performed best on Writing, since this is the highest z-score.

2.121 (a) The range is 6662-445 = 6217 and the interquartile range is $IQR = 2106 - 1334 = 772$.

(b) The maximum of 6662 is clearly an outlier and we expect it to pull the mean above the median. Since the median is 1667, the mean should be larger than 1667, but not too much larger. The mean of this data set is 1796.

(c) The best estimate of the standard deviation is 680. We see from the five number summary that about 50% of the data is within roughly 400 of the median, so the standard deviation is definitely bigger than 200. The two values above 680 would be way too large to give an estimated distance of the data values from the mean, so the only reasonable answer is 680. The actual standard deviation is 680.3.

2.123 A bell-shaped distribution with mean 3 and standard deviation 1.

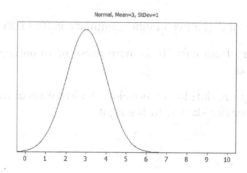

2.125 A bell-shaped distribution with mean 5 and standard deviation 2.

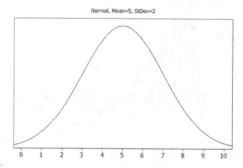

2.127 Using technology, we see that the mean rating is 51.708, the standard deviation is 26.821, and the five number summary is (0, 28, 52, 75, 99).

2.129 (a) One half of the data should have a range of 10 and all of the data should have a range of 100. The data is very bunched in the middle, with long tails on the sides. One possible histogram is shown.

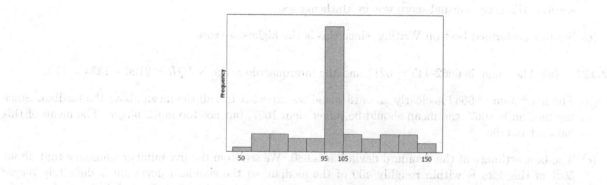

(b) One half of the data should have a range of 40 and all of the data should have a range of 50. This is a bit tricky – it means the outside 50% of the data fits in only 10 units, so the data is actually clumped on the outside margins. One possible histogram is shown.

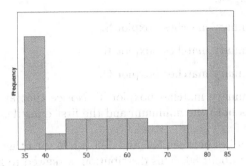

Section 2.4 Solutions

2.131 We match the five number summary with the maximum, first quartile, median, third quartile, and maximum shown in the boxplot.

(a) This five number summary matches boxplot S.

(b) This five number summary matches boxplot R.

(c) This five number summary matches boxplot Q.

(d) This five number summary matches boxplot T. Notice that at least 25% of the data is exactly the number 12, since 12 is both the minimum and the first quartile.

2.133   (a) Half of the data lies between 585 and 595, while the other half (the left tail) is stretched all the way from 585 down to about 50. This distribution is skewed to the left.

(b) There are 3 low outliers.

(c) We see that the median is at about 585 and the distribution is skewed left, so the mean is less than the median. A reasonable estimate for the mean is about 575 or 580.

2.135   (a) This distribution looks very symmetric.

(b) Since there are no asterisks on the graph, there are no outliers.

(c) We see that the median is at approximately 135. Since the distribution is symmetric, we expect the mean to be very close to the median, so we estimate the mean to be about 135.

2.137   (a) We see that $Q_1 = 260$ and $Q_3 = 300$ so the interquartile range is IQR $= 300 - 260 = 40$. We compute

$$Q_1 - 1.5(IQR) = 260 - 1.5(40) = 200,$$

and

$$Q_3 + 1.5(IQR) = 300 + 1.5(40) = 360.$$

Since the minimum (210) and maximum (320) values lie inside these values, there are no outliers.

(b) Boxplot:

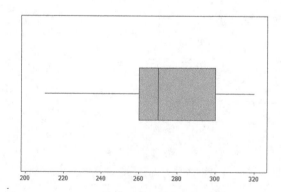

2.139 (a) We see that $Q_1 = 72$ and $Q_3 = 80$ so the interquartile range is IQR $= 80 - 72 = 8$. We compute

$$Q_1 - 1.5(IQR) = 72 - 1.5(8) = 60,$$

and

$$Q_3 + 1.5(IQR) = 80 + 1.5(8) = 92.$$

There are four data values that fall outside these values, one on the low side and three on the high side. We see that 42 is a low outlier and 95, 96, and 99 are all high outliers.

(b) Notice that the line on the left of the boxplot extends down to 63, the smallest data value that is not an outlier, while the line on the right extends up to 89, the largest data value that is not an outlier.

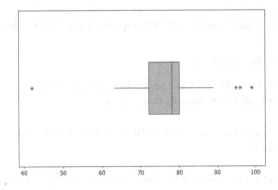

2.141 (a) The distribution is left-skewed.

(b) Half of all literacy rates are between about 73% and 98%.

(c) The five number summary is about (15, 73, 92, 98, 100).

(d) The true median is most likely lower than that shown here, because data is more likely to be available for developed countries, which have higher literacy rates.

2.143 (a) The explanatory variable is the group the person is in, and it is categorical. The response variable is hippocampus volume, and it is quantitative.

(b) The control group of people who never played football appears to have the largest hippocampal volume, while the football players with a history of concussion appears to have the smallest.

(c) Yes, there are two outliers (one high and one low) in the group of football players with a history of concussion.

(d) The third quartile appears to be at about 7000 μL.

(e) Yes, there is a quite obvious association, with those playing football having smaller brain hippocampus volume and those playing football with concussions having even smaller volume.

(f) No, we cannot conclude causation as the data come from an observational study and not an experiment. There are many possible confounding variables.

2.145 (a) Most of the data is between 0 and 500, and then the data stretches way out to the right to some very large outliers. This data is very much skewed to the right.

(b) The data appear to range from about 0 to about 10,000, so the range is about $10,000 - 0 = 10,000$.

(c) The median appears to be about 250. About half of all movies recover less than 250% of their budget, and half recover more than 250% of the budget.

(d) The very large outliers will pull the mean up, so we expect the mean to be larger than the median. (In fact, the median is 254.8 while the mean is 384.6.)

2.147 (a) Action movies appear to have the largest budgets, while horror and drama movies appear to have the smallest budgets.

(b) Action movies have by far the biggest spread in the budgets, with dramas appearing to have the smallest spread.

(c) Yes, there definitely appears to be an association between genre and budgets, with action movies having substantially larger budgets than the other three types.

2.149 (a) The lowest level of physical activity appears to be in the South, and the highest appears to be in the West.

(b) There are no outliers in any of the regions.

(c) Yes, the boxplots are very different between the different regions. Almost all of the values in the West are larger than almost all the values in the South.

2.151 The side-by-side boxplots are almost identical. Vitamin use appears to have no effect on the concentration of retinol in the blood.

2.153 (a) The five number summaries for Individual (12, 31, 39.5, 45.5, 59) and Split (22, 40, 46.5, 61, 81) show that costs tend to be higher when subjects are splitting the bill. This is also true for the means ($\overline{x}_I = 37.29$ vs $\overline{x}_S = 50.92$), while the standard deviations and IQRs show slightly more variability for those splitting the bill ($s_I = 12.54$ and $IQR_I = 14.5$ vs $s_S = 14.33$ and $IQR_S = 21$).

(b) Side-by-side plot show the distributions of costs tend to be relatively symmetric for both groups, but generally higher and slightly more variable for those splitting the bill.

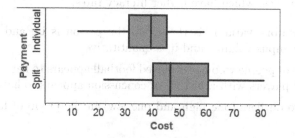

2.155 (a) The explanatory variable is whether the traffic lights are on a fixed or flexible system. This variable is categorical. The response variable is the delay time, in seconds, which is quantitative.

(b) Using technology we find the mean and standard deviation for each sample:
Timed: $\overline{x}_T = 105$ seconds and $s_T = 14.1$ seconds
Flexible: $\overline{x}_F = 44$ seconds and $s_F = 3.4$ seconds
This shows that the mean delay time is much less, 61 seconds or more than a full minute, with the flexible light system. We also see that the variability is much smaller with the flexible system.

(c) For the differences we have $\overline{x}_D = 61$ seconds and $s_D = 15.2$ seconds.

(d) The boxplot is shown. We see that there are 3 large outliers. Since this is a boxplot of the differences, this means there were three simulation runs where the flexible system *really* improved the time.

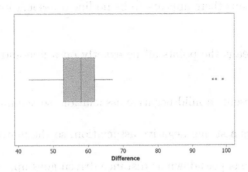

2.157 Here is one possible graph of the side-by-side boxplots:

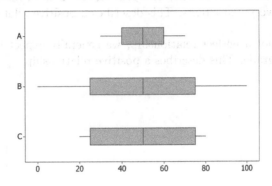

2.159 Answers will vary.

Section 2.5 Solutions

2.161 A correlation of 0 means there appears to be no linear association in the scatterplot, so the matching scatterplot is (c).

2.163 A correlation of 1 means the points all lie exactly on a line and there is a positive association. The matching scatterplot is (a).

2.165 The correlation represents a mild negative association, so the matching scatterplot is (a)

2.167 The correlation shows a strong negative association, so the matching scatterplot is (b).

2.169 Since the amount of gas goes down as distance driven goes up, we expect a negative association.

2.171 Usually someone who sends lots of texts also gets lots of them in return, and someone who does not text very often does not get very many texts. This describes a positive relationship.

2.173 While it is certainly not a perfect relationship, we generally expect that more time spent studying will result in a higher exam grade. This describes a positive relationship.

2.175 See the figure below.

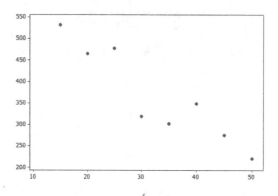

2.177 The correlation is $r = -0.932$.

2.179 (a) More nurturing is associated with larger hippocampus size, so this is a positive association.

  (b) Larger hippocampus size is associated with more resiliency, so this is a positive association.

  (c) An experiment would involve randomly assigning some children to get lots of nurturing while randomly assigning some other children to get less nurturing. After many years, we would measure the size of the hippocampus in their brains. It is clearly not ethical to assign some children to not get nurtured!

  (d) We cannot conclude that there is a cause and effect relationship in humans. No experiment has been done and there are many possible confounding variables. We can, however, conclude that there is a cause and effect relationship in animals, since the animal results come from experiments. This causation in animals probably increases the likelihood that there is a causation effect in humans as well.

2.181 (a) BMI, cortisol level, depression score, and heart rate are all positively correlated with social jetlag, while weekday hours of sleep and physical activity are negatively correlated.

(b) No, we cannot conclude causation since this is an observational study and not an experiment.

2.183 (a) A positive associate would indicate that teams that win more in the pre-season tend to win more in the regular season, while a negative association would indicate that teams that win more in the pre-season tend to lose more in the regular season.

(b) This is a very small positive correlation, so it tells you there is either no relationship or a very weak positive linear relationship between these two variables.

2.185 Both variables in the study are categorical, not quantitative. Correlation is a statistical measure of the linear relationship between two quantitative variables.

2.187 (a) The dots go up as we move left to right, so there appears to be a positive relationship. In context, that means that as a country's residents use more of the planet's resources, they tend to be happier and healthier.

(b) The bottom left is an area with low happiness and low ecological footprint, so they are countries whose residents are not very happy and don't use many of the planet's resources.

(c) Costa Rica is the highest dot – a black dot with an ecological footprint of about 2.0.

(d) For ecological footprints between 0 and 6, there is a strong positive relationship. For ecological footprints between 6 and 10, however, there does not seem to be any relationship. Using more resources appears to improve happiness up to a point but not beyond that.

(e) Countries in the top left are high on the happiness scale but are relatively low on resource use.

(f) There are many possible observations one could make, such as that countries in Sub-Saharan Africa are low on the happiness and well-being scale and also low on the use of resources, while Western Nations are high on happiness but also very high on the use of the planet's resources.

(g) For those in the bottom left (such as many countries in Sub-Saharan Africa), efforts should be devoted to improving the well-being of the people. For those in the top right (such as many Western nations), efforts should be devoted to reducing the use of the planet's resources.

2.189 (a) There are three variables mentioned: how closed a person's body language is, level of stress hormone in the body, and how powerful the person felt. Since results are recorded on numerical scales that represent a range for body language and powerful, all three variables are quantitative.

(b) People with a more closed posture (low values on the scale) tended to have higher levels of stress hormones, so there appears to be a negative relationship. If the scale for posture had been reversed, the answer would be the opposite. A positive or negative relationship can depend on how the data is recorded.

(c) People with a more closed posture (low values on the scale) tended to feel less powerful (low values on that scale), so there appears to be a positive relationship. If both scales were reversed, the answer would not change. If only one of the scales was reversed, the answer would change.

2.191 (a) A positive relationship would imply that a student who exercises lots also watches lots of television, and a student who doesn't exercise also doesn't watch much TV. A negative relationship implies that students who exercise lots tend to not watch much TV and students who watch lots of TV tend to not exercise much.

(b) A student in the top left exercises lots and watches very little television. A student in the top right spends lots of hours exercising and also spends lots of hours watching television. (Notice that there are no students in this portion of the scatterplot.) A student in the bottom left does not spend much time either exercising or watching television. (Notice that there are lots of students in this corner.) A student in the bottom right watches lots of television and doesn't exercise very much.

(c) The outlier on the right watches a great deal of television and spends very little time exercising. The outlier on the top spends a great deal of time exercising and watches almost no television.

(d) There is essentially no linear relationship between the number of hours spent exercising and the number of hours spent watching television.

2.193 (a) Men who are age 20 rate women who are the same age as they are the most attractive. This is the only age for which this is true. Men of all other ages rate younger women as more attractive.

(b) Men of all ages rate women who are in their early 20s as the most attractive.

(c) There is very little association between these variables, so the best answer is 0. (The actual correlation is 0.287.)

2.195 (a) We see in the scatterplot that the relationship is positive. This makes sense for irises: petals which are long are generally wider also.

(b) There is a relatively strong linear relationship between these variables.

(c) The correlation is positive and close to, but not equal to, 1. A reasonable estimate is $r \approx 0.9$.

(d) There are no obvious outliers.

(e) The width of that iris appears to be about 11 mm.

(f) There are two obvious clumps in the scatterplot that probably correspond to different species of iris. One type has significantly smaller petals than the other(s).

2.197 (a) See the figure.

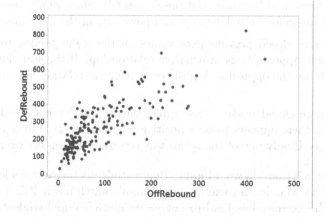

(b) There is a relatively strong positive relationship, which means players who have lots of defensive rebounds also tend to have lots of offensive rebounds. This makes sense since players who are very tall tend to get lots of rebounds in either end of the court.

(c) There are two outliers in the upper right with unusually high numbers of rebounds. We see in the data file that these two players are DeAndre Jordan who has the most defensive rebounds (829) and Andre Drummond who has the most offensive rebounds (437). Both players have more than 1100 total rebounds.

(d) We use technology to see that the correlation is 0.800. This correlation matches the strong positive linear relationship we see in the scatterplot.

2.199 Answers will vary

Section 2.6 Solutions

2.201   (a) The predicted value for the data point is $\widehat{BAC} = -0.0127 + 0.018(3) = 0.0413$. The residual is $0.08 - 0.0413 = 0.0387$. This individual's BAC was 0.0387 higher than predicted.

   (b) The slope of 0.018 tells us the expected change in BAC given a one drink increase in drinks. We expect one drink by this individual to increase BAC by 0.018.

   (c) The intercept of -0.0127 tells us that the BAC of someone who has consumed no drinks is negative. The context makes sense, but a negative BAC is not possible!

2.203   (a) The predicted value for the data point is $\widehat{Grade} = 41.0 + 3.8(10) = 79$. The residual is $81 - 79 = 2$. This student did two points better than predicted.

   (b) The slope 3.8 tells the expected change in Grade given a one hour increase in Study. We expect the grade to go up by 3.8 for every additional hour spent studying.

   (c) The intercept 41.0 tells the Grade when Study is 0. The expected grade is 41 if the student does not study at all. This context makes sense.

2.205 The regression equation is $\hat{Y} = 47.267 + 1.843X$.

2.207 The regression equation is $\hat{Y} = 641.62 - 8.42X$.

2.209   (a) The explanatory variable is the duration of the waggle dance. We use it to predict the response variable which is the distance to the source.

   (b) Yes, there is a very strong positive linear trend in the data.

   (c) We use technology to see that the correlation is $r = 0.994$. These honeybees are very precise with their timing!

   (d) We use technology to see that the regression line is $\widehat{Distance} = -399 + 1174 \cdot Duration$.

   (e) The slope is 1174, and indicates that the distance to the source goes up by 1174 meters if the dance lasts one more second.

   (f) If the dance lasts 1 second, we predict that the source is $\widehat{Distance} = -399 + 1174(1) = 775$ meters away. If the dance lasts 3 seconds, we predict that the source is $\widehat{Distance} = -399 + 1174(3) = 3123$ meters away.

2.211   (a) This is a negative association since increases in elevation are associated with decreases in cancer incidence.

   (b) The sentence is telling us the slope of the regression line.

   (c) The explanatory variable is elevation and the response variable is lung cancer incidence.

2.213   (a) For this case, the number of years playing football is 18, predicted brain size appears to be about 2700, actual brain size appears to be about 2400, and the residual is $2400 - 2700 = -300$.

   (b) The largest positive residual corresponds to the point farthest above the line. For this point, the number of years playing football appears to be 12, predicted brain size appears to be about 3000, actual brain size appears to be about 3900, so the residual is $3900 - 3000 = 900$.

   (c) The largest negative residual corresponds to the point farthest below the line. For this point, the number of years playing football appears to be 13, predicted brain size appears to be about 2900, actual brain size appears to be about 2200, so the residual is $2200 - 2900 = -700$.

2.215 The slope is 0.839. The slope tells us the expected change in the response variable (Margin) given a one unit increase in the predictor variable (Approval). In this case, we expect the margin of victory to go up by 0.839 if the approval rating goes up by 1.

The y-intercept is -36.76. This intercept tell us the expected value of the response variable (Margin) when the predictor variable (Approval) is zero. In this case, we expect the margin of victory to be -36.76 if the approval rating is 0. In other words, if *no one* approves of the job the president is doing, the president will lose in a landslide. This is not surprising!

2.217 (a) The explanatory variable is pre-season wins, the response variable is regular season wins.

(b) For a team that won 2 games in the pre-season, the predicted number of wins is $7.5 + 0.2(2) = 7.9$ wins.

(c) The slope is 0.2, which implies that for every 1 pre-season win we predict 0.2 more regular season wins.

(d) The intercept is 7.5, which implies that we predict a team with 0 pre-season wins will win 7.5 regular season games. This is a reasonable assumption and prediction.

(e) The regression line predicts 27.5 wins for a team that wins 100 pre-season games, which is definitely not appropriate because we are extrapolating well away from the possible number of pre-season wins. (There are only four pre-season games.)

2.219 The man with the largest positive residual weighs about 190 pounds and has a body fat percentage about 40%. The predicted body fat percent for this man is about 20% so the residual is about $40 - 20 = 20$.

2.221 (a) For 35 cm, the predicted body fat percent is $\widehat{BodyFat} = -47.9 + 1.75 \cdot (35) = 13.35\%$. For a neck circumference of 40 cm, the predicted body fat percent is $\widehat{BodyFat} = -47.9 + 1.75 \cdot (40) = 22.1\%$.

(b) The slope of 1.75 indicates that as neck circumference goes up by 1 cm, body fat percent goes up by 1.75.

(c) The predicted body fat percent for this man is $\widehat{BodyFat} = -47.9 + 1.75 \cdot (38.7) = 19.825\%$, so the residual is $11.3 - 19.825 = -8.525$.

2.223 (a) See the figure. There is a clear linear trend, so it is reasonable to construct a regression line.

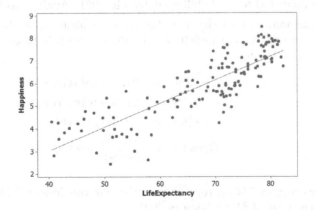

(b) Using technology, we see that the regression line is $\widehat{Happiness} = -1.09 + 0.103 \cdot LifeExpectancy$.

(c) The slope of 0.103 indicates that for an additional year of life expectancy, the happiness rating goes up by 0.103.

2.225 (a) The three variables are: average growth rate over the decade, predicted 2021 debt-to-GDP ratio, and predicted 2021 deficit.

(b) We have $\widehat{Ratio} = 129 - 19.1(Growth)$.

(c) The slope indicates that as the growth rate goes up by 1%, the predicted 2021 debt-to-GDP ratio goes down by 19.1%. This is a very impressive and good effect! The intercept of 129 indicates that if the growth rate is 0% (indicating no economic growth over the decade), the predicted 2021 debt-to-GDP ratio is 129%, which would be very bad for the country.

(d) Since $\widehat{Ratio} = 129 - 19.1(2) = 90.8$, we see that at a growth rate of 2%, the debt-to-GDP ratio is predicted to be 90.8% in 2021. Since $\widehat{Ratio} = 129 - 19.1(4) = 52.6$, we see that at a growth rate of 4%, the debt-to-GDP ratio is predicted to be 52.6% in 2021. This is a very big difference!

(e) We see from the answers to part (d) that the answer is likely to be slightly bigger than 2%. To find the actual value, we substitute a Ratio of 90% into the regression line and solve for the growth rate:

$$\begin{aligned} \widehat{Ratio} &= 129 - 19.1(Growth) \\ 90 &= 129 - 19.1(Growth) \\ -39 &= -19.1(Growth) \\ Growth &= \frac{-39}{-19.1} = 2.04. \end{aligned}$$

If the growth rate averages 2.04% over the decade 2011 to 2021, we expect the ratio to hit 90% in 2021.

(f) We have $\widehat{Deficit} = 2765 - 680(Growth)$.

(g) The slope indicates that as the growth rate goes up by 1%, the predicted deficit in 2021 goes down by 680 billion dollars. This is again a very impressive and good effect! The intercept of 2765 indicates that if the growth rate is 0% (indicating no economic growth), the predicted 2021 deficit is 2765 billion dollars, or 2.765 trillion dollars. This would not be a good outcome.

(h) Since $\widehat{Deficit} = 2765 - 680(2) = 1405$, we see that at a growth rate of 2%, the deficit is predicted to be 1405 billion dollars in 2021. Since $\widehat{Deficit} = 2765 - 680(4) = 45$, we see that at a growth rate of 4%, the deficit is predicted to be 45 billion dollars in 2021. Again, this is a very big difference!

(i) We see from the answers to part (h) that the answer is likely to be slightly bigger than 2%. To find the actual value, we substitute a Deficit of 1.4 trillion (which is 1400 billion) into the regression line and solve for the growth rate:

$$\begin{aligned} \widehat{Deficit} &= 2765 - 680(Growth) \\ 1400 &= 2765 - 680(Growth) \\ -1365 &= -680(Growth) \\ Growth &= \frac{-1365}{-680} = 2.007. \end{aligned}$$

If the growth rate averages 2.007% (or just over 2%) over the decade 2011 to 2021, we expect the deficit to be at the current level of $1.4 trillion in 2021.

Section 2.7 Solutions

2.227 (a) The variable *Happiness* is quantitative, the variable *Footprint* is quantitative, and the variable *Region* is categorical.

(b) Regions 1 (Latin America) and 2 (Western nations) seem to have the greatest happiness score. Region 2 (Western nations) seems to have the greatest ecological footprint.

(c) Region 4 (Sub-Saharan Africa) seems to have the lowest happiness scores. The ecological footprint also tends to be low in that region.

(d) Yes, overall, having a greater ecological footprint appears to be associated with having a higher happiness score.

(e) No. Considering only those countries in region 2 (Western nations), there does not seem to be a relationship between happiness and footprint. A greater ecological footprint does not appear to be associated with a higher happiness score.

(f) It is in the top left, with a relatively large happiness score and a relatively small ecological footprint.

(g) For countries in region 4, we should focus on trying to increase the happiness score. For countries in region 2, we should focus on decreasing the ecological footprint.

2.229 (a) There are three variables displayed in the figure. Hippocampus size is quantitative and number of years playing football is quantitative, while the group (control, football with concussion, football without concussion) is categorical.

(b) The blue dots represent the study participants who are in the control group and have never played football, so the "years of football" for all these participants is zero.

(c) The general trend in the scatterplot shows a negative association between years playing football and hippocampus size.

(d) The reddish brown line (for the football with concussion group) is lower, which tells us that football players with a history of concussions appear to have smaller hippocampus sizes.

(e) The green line (for football players without a history of concussions) has the steeper slope.

2.231 (a) While it varies each year, the general trend for both positions is running the 40 yard dash in less time, so faster.

(b) Cornerbacks were slightly faster.

(c) No, from year to year it varies which position runs faster on average.

2.233 (a) There are 7 purple dots at the 24-minute mark, so they were losing in 7 games at halftime.

(b) There are 2 purple dots at the 0-minute mark, so they only lost two of these games (both times by more than 13 points.)

2.235 (a) In 1970 the majority of people living in extreme poverty were from Asia.

(b) In 2015 the majority of people living in extreme poverty were from Africa.

(c) In 1970, the distribution has two peaks (it is *bimodal*), with one peak below \$1/day and another between \$10/day and \$20/day. This clearly distinguishes the "third world" from the "first world". In 2015, the overall distribution is more symmetric and bell-shaped on the log-scale provided.

2.237 (a) There are 6 different categories shown, from "<10%" to "≥30%".

(b) In 1990, it appears that the highest category needed is "10%-14%". In 2000, the highest category needed is "20%-24%". In 2010, many states are in the top category "≥30%".

2.239 (a) We see in the spaghetti plot that for *every* state, the percent obese more than doubled during this time period.

(b) There is more variability in 2014.

(c) In 1990, the state with the largest percent obese is Mississippi, with 15.0% obese. In 2014, the state with the smallest percent obese is Colorado, with 21.3% obese.

2.241 (a) We see that "bro" is most commonly used in Texas and surrounding states.

(b) We see that "buddy" is most commonly used in the midwest.

2.243 (a) The eastern half of the country is much more heavily populated than the western half.

(b) Answers will vary

2.245 (a) We can see that there are two distinct clusters of dots, with one cluster having smaller petals (in length and width) than the other.

(b) Setosa has the smallest petals, while Virginica has the largest petals.

2.247 Sepal length is on the vertical axis, and the green dots are highest up.

2.249 (a) Whites are decreasing the most in terms of percentage of the US population, from 85% in 1960 to (projected) 43% in 2060.

(b) Hispanics are increasing the most in terms of percentage of the US population, from 4% in 1960 to (projected) 31% in 2060.

(c) Blacks are staying the most constant in terms of percentage of the US population, from 10% in 1960 to (projected) 13% in 2060.

2.251 (a) The commutes on the carbon bike tend to be longer in distance than the commutes on the steel bike.

(b) The carbon bike is slightly faster, on average, but also tends to cover more distance on the commute, making the overall trip longer in minutes. Perhaps the devices measuring distance, time, or speed give slightly different readings between the two bikes.

(c) Distance is associated with both type of bike and commute time in minutes, confounding the relationship between the two. Distance is a confounding variable.

(d) To minimize commute time, Dr. Groves should ride the carbon bike (which has a slightly faster average speed), but take the shorter distance commutes that are typically taken with the steel bike in this dataset. He might also want to check the accuracy of the recording devices used with each bike.

2.253 (a) In 2014, New England had the highest support for same-sex marriage.

(b) In 2014, South Central had the lowest support for same-sex marriage.

(c) The Mountain states displayed the smallest increase (about 9%), and the Midwest displayed the largest (about 24%, although New England and Pacific are close at 23%).

(d) This data could also have been visualized with a spaghetti plot, in which each time series would be shown on the same plot, or with a dynamic heat map, in which a map of the US could have been shown color-coded according to the level of support, with the colors changing over time.

2.255 Answers will vary.

2.257 There is a strong tendency for blacks to live near other blacks in one region of St. Louis (near the middle of the city), and for whites to live near other whites.

2.259 (a) Early morning (6 am to 7 am) has the highest percent cloud cover in August.

(b) Summer (around August) tends to be the least windy season for Chicago.

2.261 (a) This video is conveying the endangered status and variety of gazelle species.

(b) Answers will vary.

2.263 Answers will vary.

2.265 Answer will vary.

2.267 Answers will vary.

2.269 Answers will vary.

2.271 Answers will vary.

Unit A: Essential Synthesis Solutions

A.1 (a) Yes

(b) No

A.3 (a) No

(b) Yes

A.5 (a) No

(b) No

A.7 (a) One categorical variable

(b) Bar chart or pie chart

(c) Frequency or relative frequency table, proportion

A.9 (a) One categorical variable and one quantitative variable

(b) Side-by-side boxplots, dotplots, or histograms

(c) Statistics by group or difference in means

A.11 (a) Two categorical variables

(b) Segmented or side-by-side bar charts

(c) Two-way table or difference in proportions

A.13 (a) One quantitative variable

(b) Histogram, dotplot, or boxplot

(c) Mean, median, standard deviation, range, IQR

A.15 (a) Two quantitative variables

(b) Scatterplot

(c) Correlation or slope from regression

A.17 (a) This is an experiment since subjects were randomly assigned to one of three groups which determined what method was used.

(b) This study could not be "blind" since both the participants and those recording the results could see what each had applied.

(c) The sample is the 46 subjects participating in the experiment. The intended population is probably anyone who might consider using black grease under the eyes to cut down on glare from the sun.

(d) One variable is the improvement in contrast sensitivity, and this is a quantitative variable. A second variable records what group the individual is in, and this is a categorical variable.

(e) Since we are examining the relationship between a categorical variable and a quantitative variable, we could use side-by-side boxplots to display the results.

A.19 (a) The cases are the students (or the students' computers). The sample size is $n = 45$. The sample is not random since the students were specifically recruited for the study.

(b) This is an observational study, since none of the variables was actively manipulated.

(c) For each student, the variables recorded are: number of active windows per lecture, percent of windows that are distracting, percent of time on distracting windows, and score on the test of the material. All of these variables are quantitative. (Note whether or not a window is distracting is categorical, but for each student the *percentage* of distracting windows is quantitative.)

(d) The number of active windows opened per lecture is a single quantitative variable, so we might use a histogram, dotplot, or boxplot. If we want outliers clearly displayed, a boxplot would be the best choice.

(e) The association described is between two quantitative variables, so we would use a scatterplot. An appropriate statistic would be a correlation. Since more time on distracting websites is associated with lower test scores, it is a negative association.

(f) No, we cannot conclude that the time spent at distracting sites causes lower test scores, since this is an observational study not an experiment. There are many possible confounding variables and we cannot infer a cause and effect relationship (although there might be one).

(g) We consider the time on distracting websites the explanatory variable, and the exam score the response variable.

(h) To make this cause and effect conclusion, we would need to do a randomized experiment. One option would be to randomly divide a group of students into two groups and have one group use distracting websites and the other not have access to such websites. Compare test scores of the two groups at the end of the study. It is probably not feasible to require one group to visit distracting websites during class!

A.21 (a) The sample is the 86 patients in the study. The intended population is all people with bladder cancer.

(b) One variable records whether or not the tumor recurs, and one records which treatment group the patient is in. Both variables are categorical.

(c) This is an experiment since treatments were randomly assigned. Since the experiment was double-blind, we know that neither the participants nor the doctors checking for new tumors knew who was getting which treatment.

(d) Since we are looking at a relationship between two categorical variables, it is reasonable to use a two-way table to display the data. The categories for treatment are "Placebo" and "Thiotepa" and the categories for outcome are "Recurrence" and "No Recurrence". The two-way table is shown.

	Recurrence	No recurrence	Total
Placebo	29	19	48
Thiotepa	18	20	38
Total	47	39	86

(e) We compare the proportion of patients for whom tumors returned between the two groups. For the placebo group, the tumor returned in $29/48 = 0.604 = 60.4\%$ of the patients. For the group taking the active drug, the tumor returned in $18/38 = 0.474 = 47.4\%$ of the patients. The drug appears to be more effective than the placebo, since the rate of recurrence of the tumors is lower for the patients in the thiopeta treatment group than for patients in the placebo group.

A.23 (a) The cases are the bills in this restaurant. The sample size is $n = 157$.

  (b) There are seven variables. *Bill*, *Tip*, and *PctTip* are quantitative, while *Credit*, *Server*, and *Day* are categorical. The seventh variable *Guests* could be classified as either quantitative or categorical depending on what we wanted to do with it.

  (c) The mean is $\bar{x} = 16.62$ and the standard deviation is $s = 4.39$. The five number summary is $(6.7, 14.3, 16.2, 18.2, 42.2)$. We use the $1.5 \cdot IQR$ rule to find how large or small a tip percentage has to be to qualify as an outlier. We have $IQR = 18.2 - 14.3 = 3.9$ and we compute

$$Q_1 - 1.5 \cdot IQR = 14.3 - 1.5(3.9) = 8.45$$
$$Q_3 + 1.5 \cdot IQR = 18.2 + 1.5(3.9) = 24.05.$$

  Any data value less than 8.45 or greater than 24.05 is an outlier. Looking at the minimum (6.7) and maximum (42.2) in the five number summary, it is obvious that there are both small and large outliers in this data set.

  (d) Other than several large outliers, the histogram is symmetric and bell-shaped.

  (e) A table of type of payment and day of the week is shown. The proportion of bills paid with a credit card on Thursday is $\hat{p}_{th} = 12/36 = 0.333$ and the percent of bills paid with a credit card on Friday is $\hat{p}_f = 4/26 = 0.154$. These two proportions are quite different, so there appears to be an association between these two variables. A much larger percentage appear to pay with a credit card on Thursday than on Friday, perhaps because some people have run out of cash on Thursday but then are flush with cash again after getting paid on Friday.

	m	t	w	th	f	Total
n (cash)	14	5	41	24	22	106
y (credit)	6	8	21	12	4	51
Total	20	13	62	36	26	157

  (f) *PctTip* is a quantitative variable and *Server* is categorical, so we might use side-by-side boxplots. See the graph below. We see that Server A appears to get the highest median tip percentage (and has two large outliers) while Server B and C are similar.

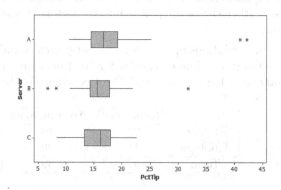

  (g) The explanatory variable is *Bill* while the response variable is *PctTip*. See the scatterplot. There are several large outliers with high values for *PctTip* – percentages between 40% and 45%. There is also a possible outlier on the right, with a value of about 70 for *Bill*. The relationship is not strong at all and may be slightly positive.

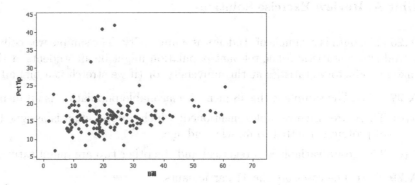

(h) The correlation is $r = 0.135$

Unit A: Review Exercise Solutions

A.25 The data is a sample of students at a university. The sample was collected by surveying students taking introductory statistics. The relevant population might be all students at this university, or all students who take introductory statistics at this university, or (if we stretch it a bit) all university students.

A.27   (a) The sample is the 48 men. A reasonable population is all men.

  (b) There are three variables mentioned: which group a man is assigned to (exercise or rest), the amount of protein converted to muscle, and age.

  (c) The group variable is categorical and the other two are quantitative.

A.29   (a) The cases are the 41 participants.

  (b) There are many variables in this study. The only categorical variable is whether or not the person participated in the meditation program. All other variables are quantitative variables. These variables include (at minimum):

- Brain wave activity before
- Brain wave activity after
- Brain wave activity 4 months later
- Immune response after 1 month
- Immune response after 2 months
- Negative survey before
- Negative survey after
- Positive survey before
- Positive survey after

  (c) The explanatory variable is whether or not the person participated in the meditation program.

  (d) The data set will have 41 rows (one for each participant) and at least 10 columns (one for each variable).

A.31 The sample is the fish in that one day's catch. The population is all fish in that area.

A.33 No, we cannot conclude that about 79% of all people think physical beauty matters, since this was a volunteer sample in which only people who decided to vote were included in the sample, and only people looking at *cnn.com* even had the opportunity to vote. The sample is the 38,485 people who voted. The population if we made such an incorrect conclusion would be all people. There is potential for sampling bias in every volunteer sample.

A.35   (a) Number the rows from 1 to 100 and the plants within each row from 1 to 300. Use a computer random number generator to pick a number between 1 and 100 to select a row and a second number between 1 and 300 to identify the plant within that row. Repeat until 30 different plants have been selected. Other options are possible: for example, we could number the plants from 1 to 3000 and randomly select 30 numbers between 1 and 3000.

  (b) Here is the start of one sample.

```
Row     Plant
#94     #180
#83     # 81
#10     #222
```

Other options are possible.

A.37 (a) One possibility: Find professors who give an easy first quiz and professors who give a hard first quiz. Compare their students' grades on a common exam.

  (b) The teaching styles of the professors might be a confounding factor. If professors choose to give either an easy first quiz or a hard first quiz, there are probably other differences in the teaching styles which could also dramatically impact the grades on the exam.

  (c) Randomly divide students into two groups and give one group an easy first quiz and the other group a hard first quiz. Keep everything else as similar as possible between the two groups, such as professors, other quizzes, homework, etc. Compare grades on the common exam.

A.39 Snow falls when it is cold out and the heating plant will be used more on cold days than on warm days. Also, when snow falls, people have to shovel the snow and that can lead to back pain. Notice that the confounding variable has an association with *both* the variables of interest.

A.41 (a) The cases are university students. One variable is whether the student lives in a single-sex or co-ed dorm. This is a categorical variable. The other variable is how often the student reports hooking up for casual sex, which is quantitative.

  (b) The type of dorm is the explanatory variable and the number of hook-ups is the response variable.

  (c) Yes, apparently the studies show that students in same sex dorms hook-up for casual sex more often, so there is an association.

  (d) Yes, the president is assuming that there is a causal relationship, since he states that "single sex dorms *reduce* the number of student hook-ups".

  (e) There is no indication that any variable was manipulated, so the studies are probably observational studies.

  (f) The type of student who requests a single sex dorm might be different from the type of student who requests a co-ed dorm. There are other possible confounding variables.

  (g) No! We should not assume causation from an observational study.

  (h) He is assuming causation when there may really only be association.

A.43 (a) The cases are the students and the sample size is 70.

  (b) There are four variables mentioned. One is the treatment group (walk in sync, walk out of sync, or walk any way). The other three are the quantitative ratings given on the three questions of closeness, liking, and similarity.

  (c) This is an experiment since the students were actively told how to follow the accomplice.

  (d) There are many possible ways to draw this; one is shown in the figure.

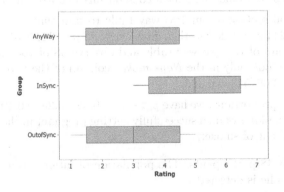

(e) One possible graph to use to look at a relationship of number of pill bugs killed by which treatment group the student was in would be a side-by-side boxplot, since one of these variables is quantitative and one is categorical. We could also use comparative dotplots or comparative histograms.
We would use a scatterplot to look at the association of number of pill bugs killed with the rating given on the liking accomplice scale, since both of these are quantitative variables.

A.45 Answers will vary. One possible sample is shown below.

$$7692, \ 1708, \ 0099, \ 4755, \ 1406, \ 4937, \ 6647, \ 2496, \ 3850, \ 4673$$

See the technology notes to see how to use specific technology to select a random sample.

A.47   (a) This is an experiment since facial features were actively manipulated.

 (b) This "blinding" allows us to get more objective reactions to the video clips.

 (c) We use $\overline{x}_S$ for the mean of the smiling group and $\overline{x}_N$ for the mean of the non-smiling group. The difference in means is $\overline{x}_S - \overline{x}_N = 7.8 - 5.9 = 1.9$.

 (d) Since the results come from a randomized experiment, a substantial difference in the mean ratings would imply that smiling causes an increase in positive emotions.

A.49 The two-way table with row and column totals is shown.

	Near-death experience	No such experience	Total
Cardiac arrest	11	105	116
Other cardiac problem	16	1463	1479
Total	27	1568	1595

To compare the two groups, we compute the percent of each group that had a near-death experience. For the cardiac arrest patients, the percent is $11/116 = 0.095 = 9.5\%$. For the patients with other cardiac problems, the percent is $16/1479 = 0.011 = 1.1\%$. We see that approximately 9.5% of the cardiac arrest patients reported a near-death experience, which appears to be much higher than the 1.1% of the other patients reporting this.

A.51   (a) Since no one assigned smoking or not to the participants, this is an observational study. Because this is an observational study, we can not use this data to determine whether smoking influences one's ability to get pregnant. We can only determine whether there is an association between smoking and ability to get pregnant.

 (b) The sample collected is on women who went off birth control in order to become pregnant, so the population of interest is women who have gone off birth control in an attempt to become pregnant.

 (c) We look in the total section of our two way table to find that out of the 678 women attempting to become pregnant, 244 succeeded in their first cycle, so $\hat{p} = 244/678 = 0.36$. For smokers we look only in the *Smoker* column of the two way table and observe 38 of 135 succeeded, so $\hat{p}_s = 38/135 = 0.28$. For non-smokers we look only in the *Non-smoker* column of the two way table and observe 206 of 543 succeeded, so $\hat{p}_{ns} = 206/543 = 0.38$.

 (d) For the difference in proportions, we have $\hat{p}_{ns} - \hat{p}_s = 0.38 - 0.28 = 0.10$. This means that in this sample, the percent of non-smoking women successfully getting pregnant in the first cycle is 10 percentage points higher than the percent of smokers.

A.53   (a) The sample is 48 participants. The population of interest is all people. The variable is whether or not each person's lie is detected.

(b) The proportion of time the lie detector fails to report deception is $\hat{p} = 17/48 = 0.35$.

(c) Since the lie detector fails 35% of the time, it is probably not reasonable to use it.

A.55 The histogram is relatively symmetric and bell-shaped. The mean appears to be approximately $\bar{x} = 7$. To estimate the standard deviation, we estimate an interval centered at 7 that contains approximately 95% of the data. The interval from 3 to 11 appears to contain almost all the data. Since 3 and 11 are both 4 units from the mean of 7, we have $2s = 4$, so the standard deviation appears to be approximately $s = 2$.

A.57 (a) There appear to be some low outliers pulling the mean well below the median. Half of the growing seasons over the last 50 years have been longer than 275 days and half have been shorter. Some of the growing seasons have been extremely short and have pulled the mean down to 240 days.

(b) Here is a smooth curve that could represent this distribution.

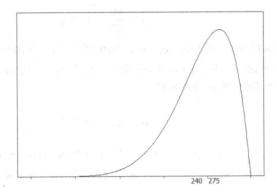

(c) The distribution is skewed to the left.

A.59 (a) The distribution has a right skew. There are a number of apparent outliers on the right side.

(b) The actual median is 140 ng/ml. Estimates between 120 and 160 are reasonable.

(c) The actual mean is 189.9 ng/ml. Estimates between 160 and 220 are reasonable. Note that the outliers and right skew should make the mean larger than the median.

A.61 The values are in order smallest to largest, and since more than half the values are 1, the median is 1. We calculate the mean to be $\bar{x} = 3.2$. In this case, the mean is probably a better value (despite the fact that 12 might be an outlier) since it allows us to see that some of the data values are above 1.

A.63 (a) Using software, we see that $\bar{x} = 0.272$ and $s = 0.237$.

(b) The largest concentration is 0.851. The z-score is

$$z\text{-score} = \frac{x - \bar{x}}{s} = \frac{0.851 - 0.272}{0.237} = 2.44.$$

The largest value is almost two and a half standard deviations above the mean, and appears to be an outlier.

(c) Using software, we see that

$$\text{Five number summary} = (0.073, 0.118, 0.158, 0.358, 0.851).$$

(d) The range is $0.851 - 0.073 = 0.778$ and the interquartile range is $IQR = 0.358 - 0.118 = 0.240$.

A.65   (a) The average for both joggers is 45, so they are the same.

(b) The averages are the same, but the set of times for jogger 1 has a much lower standard deviation.

A.67 It is helpful to compute the row and column totals for the table as shown below.

	Greater than 50%	Less than 50%	Total
Rosiglitazone	5	42	47
Placebo	21	27	48
Total	26	69	95

(a) A total of 47 patients received rosiglitazone, while 48 patients received a placebo.

(b) There were a total of 95 patients in the study, and $42 + 27 = 69$ of them had less than 50% blockage, so we have

$$\hat{p} = \frac{69}{95} = 0.726.$$

About 72.6% of the patients had blockage less than 50% after 6 months.

(c) We consider only patients with greater than 50% blockage, a total of 26 patients. Of these, only 5 patients were on rosiglitazone, so we have

$$\hat{p} = \frac{5}{26} = 0.192.$$

Only 19.2% of the patients with greater than 50% blockage were taking the drug rosiglitazone.

(d) We consider only the 48 patients given a placebo. Of these, 27 had less than 50% blockage, so we have

$$\hat{p} = \frac{27}{48} = 0.5625.$$

We see that 56.25% of the patients given a placebo had less than 50% blockage.

(e) We are comparing rosiglitazone to a placebo, which are the rows of the table, so we find the proportion with less than 50% blockage in each row. For rosiglitazone, the proportion is $42/47 = 0.894$. For those taking a placebo, the proportion is $27/48 = 0.563$. The percent is quite a bit higher for those taking the active drug, rosiglitazone.

(f) From the results of this sample, it appears that rosiglitazone is effective at limiting coronary blockage. The percent of patients whose blockage was less than 50% after 6 months was almost 90% for those taking the drug and it was only about 56% for those taking a placebo.

A.69   (a) This is an experiment. Double-blind means that neither the patients nor the doctors making the cancer diagnosis knew who was getting the drug and who was getting a placebo.

(b) There are two variables: one records the presence or absence of prostate cancer and the other records whether the individual was in the finasteride group or the placebo group.

(c) Here is a two-way table for treatment groups and cancer diagnosis.

	Cancer	No cancer	Total
Finasteride	804	3564	4368
Placebo	1145	3547	4692
Total	1949	7111	9060

(d) We have

$$\text{Percent receiving finasteride} = \frac{4368}{9060} = 48.2\%.$$

(e) A total of 1949 men were found to have cancer, and 1145 of these were in the placebo group, so we have

$$\hat{p} = \frac{1145}{1949} = 58.7\%.$$

(f) We have the following cancer rates in each group:

$$\text{Percent on finasteride getting cancer} = \frac{804}{4368} = 18.4\%$$
$$\text{Percent on a placebo getting cancer} = \frac{1145}{4692} = 24.4\%$$

The percent getting cancer appears to be quite a bit lower for those taking finasteride.

A.71 (a) We see that 22 of the 72 participants found much improvement in sleep quality, so the proportion is $22/72 = 0.306$.

(b) We combine the results of the "medication" and "both" columns to find that $5 + 10 = 15$ of the $17 + 19 = 36$ people on medication had much improvement in sleep quality, so the proportion is $15/36 = 0.417$.

(c) Thirty-four participants had no improvement and within this row we find $8 + 3 = 11$ who had medication, so the proportion is $11/34 = 0.324$.

(d) For the denominator we need the totals for the "medication" and "neither" groups who did not receive training, $17 + 18 = 35$. In the numerator we add the counts for "much" and "some" improvement in each of the two groups, $5 + 0 + 4 + 1 = 10$, so the proportion without training who received some or much improvement is $10/35 = 0.286$.

A.73 (a) The minimum values are similar, but each quartile after that and the maximum are extremely different. The birth rate distributions do appear to be different in developed and undeveloped nations, with the values for birth rates in undeveloped countries tending to be much higher.

(b) Many of the undeveloped countries would be high outliers in birth rate if they were considered developed. The general rule of thumb for detection of larger outliers is $Q_3 + 1.5(IQR)$, which for developed nations (where $IQR = 14.5 - 9.5 = 5.0$) is $14.5 + 1.5(5.0) = 22.0$, so in fact even the median for undeveloped countries would be an outlier in developed countries! By contrast, none of the developed countries would be outliers if considered undeveloped. The lower bound for outliers based on the undeveloped quartiles (where $IQR = 31.5 - 18.8 = 12.7$) is $Q_1 - 1.5 * IQR = 18.8 - 1.5 * 12.7 = -0.25$ and there are no negative birthrates.

(c) Kazakhstan' birth rate of 22.7 would be an outlier among developed countries (see the calculation above which puts the threshold at anything over 22.0). Kazakhstan would not be an outlier in the distribution of undeveloped countries. In fact its birth rate is close to the median for undeveloped countries.

(d) Sketch should be similar to the boxplots shown below (although you would need to use the raw data to determine exactly where the right whisker ends for developed countries.)

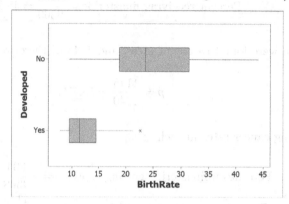

A.75 For the 13 teenage patients, the five number summary for blood pressure is (100, 104, 130, 140, 156). For the fifteen patients in their eighties, the five number summary is (80, 110, 135, 141, 190).

The range for the patients in their eighties, $190 - 80 = 110$, is quite a bit larger than for the teenage patients, $156 - 100 = 56$. However the interquartile ranges, IQR$= 140 - 104 = 36$ for the patients in their teens and IQR$=141 - 110 = 31$ for those in their 80s, are fairly similar. Thus, while we see more variability at both extremes for the older patients, the distribution of the middle 50% of the blood pressure readings are not very different between the two age groups.

The standard deviation for the blood pressure readings for the 13 ICU patients in their teens is $s = 19.57$ while the standard deviation for the 15 patients in their 80s is $s = 31.23$. This reinforces the fact that the variability is greater for patients in their 80s.

A.77　(a) We see that the interquartile range is $IQR = Q_3 - Q_1 = 149 - 15 = 134$. We compute:

$$Q_1 - 1.5(IQR) = 15 - 1.5(134) = 15 - 201 = -186.$$

and

$$Q_3 + 1.5(IQR) = 149 + 1.5(134) = 149 + 201 = 350.$$

Outliers are any values outside these fences. In this case, there are four outliers that are larger than 350. The four outliers are 402, 447, 511, and 536.

(b) A boxplot of time to infection is shown:

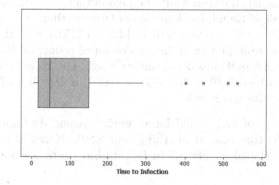

A.79 (a) The median appears to be about 500 calories higher for the males than for the females. The largest outlier of 6662 calories in one day is a male, but the females have many more outliers.

(b) Yes, there does appear to be an association. Females appear to have significantly lower calorie consumption than males. We see that every number in the five number summary is higher for males than it is for females. The median for females is even lower than the first quartile for males.

A.81 Both distributions are relatively symmetric with one or two outliers. In general, the blood pressures of patients who lived appear to be slightly higher as a group than those of the patients who died. The middle 50% box for the surviving patients is shifted to the right of the box for patients who died and shows a smaller interquartile range. Both quartiles and the median are larger for the surviving group. Note that the boxplots give no information about how many patients are in each group. From the original data table, we can find that 40 of the 200 patients died and the rest survived.

A.83 (a) Yes, it is appropriate to use the 95% rule, since we see in the histogram of blood pressures that the distribution is approximately symmetric and bell-shaped.

(b) We expect 95% of the data values to lie within two standard deviations of the mean, so we have

$$\bar{x} \pm 2s = 132.28 \pm 2(32.95) = 66.38 \text{ and } 198.18.$$

(c) There are 186 systolic values between 66.38 and 198.18 among the 200 cases in **ICUAdmissions**, or $186/200 = 93\%$ of the data values within the interval $\bar{x} \pm 2s$.

(d) This data matches the 95% rule very well.

A.85 (a) A scatterplot of verbal vs math SAT scores is shown below with the regression line (which is almost perfectly flat).

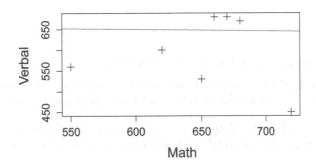

(b) Using technology we find the correlation between math and verbal scores for this sample is $r = -0.071$.

(c) Based on this small sample of seven pairs, computing a regression line to help predict verbal scores based on math scores is not very useful. The flat line in the scatterplot shows no consistent positive or negative linear trend between these variables. This is also seen with the sample correlation ($r = -0.071$) which is very close to zero.

A.87   (a) The scatterplot for the original five data points is shown below.

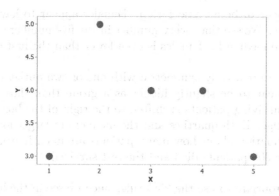

(b) We use technology to find $r = -0.189$. Both the scatterplot and the correlation show almost no linear relationship between x and y.

(c) The scatterplot with the extra point added is shown below.

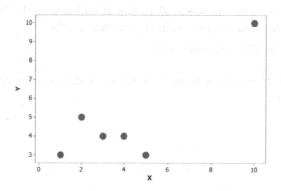

(d) The correlation (with the extra data point) is $r = 0.836$.

(e) When the outlier $(10, 10)$ is added, the correlation is suddenly very strong, and even changes from negative to positive. The outlier has a very substantial effect on the correlation.

A.89   (a) The general trend in this data appears to be up and to the right, so there is a mostly positive relationship between height and weight. This makes sense, since taller people are more likely to weigh more.

(b) Individual A is short and light in weight, individual B is tall and heavy. Individual C is relatively short and heavy, whereas individual D is relatively tall and thin.

A.91   (a) A positive association means that large values of one variable tend to be associated with large values of the other; in this case, that taller people generally weigh more. A negative association means that large values of one variable tend to be associated with small values of the other; in this case, that tall people generally weigh less than shorter people. Since we expect taller people to generally weigh more, we expect a positive relationship between these two variables.

(b) In the scatterplot, we see a positive upward relationship in the trend (as we expect) but it is not very strong. It appears to be approximately linear.

(c) The outlier in the lower right corner appears to have height about 83 inches (or 6 ft 11 inches) and weight about 135 pounds. This is a very tall thin person! (It is reasonable to suspect that this person may have entered the height incorrectly on the survey. Outliers can help us catch data-entry errors.)

A.93   (a) Here is a scatterplot of Jogger A vs Jogger B.

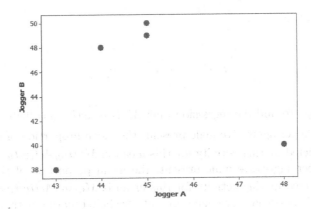

(b) The correlation between the two joggers is -0.096.

(c) The correlation between the two joggers with the windy race added is now 0.562.

(d) Adding the results from the windy day has a very strong effect on the relationship between the two joggers!

A.95   (a) We are attempting to predict rural population with land area, so land area is the explanatory variable, and percent rural is the response.

(b) There appears to be some positive correlation between these two variables, so the most likely correlation is 0.60.

(c) Using technology the regression line is: $\widehat{Rural} = 30.35 + 0.075(LandArea)$. The slope is 0.079, which means percent rural goes up by about 0.075 with each increase in 1,000 sq km of country size.

(d) The intercept does not make sense, since a country of size zero would have no population at all!

(e) The most influential country is the one in the far top right, which is Uzbekistan (UZB). This is due to the fact that Uzbekistan is much larger then any of the other countries sampled, so it appears to be an outlier for the explanatory variable.

(f) Predicting the percent rural for USA with the prediction equation gives $\widehat{Rural} = 30.35+0.075(9147.4) = 716.4$. This implies that 716% of the United States population lives in rural areas, which doesn't make any sense at all. The regression line does not work at all for the US, because we are extrapolating so far outside of the original sample of 10 land area values. The US is much larger in area than any of the countries that happened to be picked for the random sample.

A.97   (a) For the five groups with no hyper-aggressive male, we see that $\bar{x} = 0.044$ with $s = 0.062$. For the groups with a hyper-aggressive male, we have $\bar{x} = 0.390$ with $s = 0.288$. There is a large difference in the mean proportion of time females spend in hiding between the two groups.

(b) A scatterplot of female hiding proportions vs mating activity is shown below.

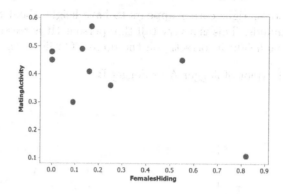

(c) We use technology to find the regression line: $\widehat{MatingActivity} = 0.49 - 0.33 \cdot FemalesHiding$.

(d) If there is not a hyper-aggressive male present, the mean proportion of time females spend in hiding is 0.044. The predicted mating activity for this mean is $\widehat{MatingActivity} = 0.49 - 0.33 \cdot (0.044) = 0.475$. If there is a hyper-aggressive male present, the mean proportion of time females spend in hiding is 0.390. The predicted mating activity for this mean is $\widehat{MatingActivity} = 0.49 - 0.33 \cdot (0.390) = 0.361$. If there is not a hyper-aggressive male present, predicted mating activity is 0.475. If there is a hyper-aggressive male present, predicted mating activity is 0.361.

(e) Don't hang out with any hyper-aggressive males!

A.99 (a) There is a strong linear positive trend, with no obvious outliers.

(b) Household income appears to be a stronger predictor and more strongly correlated.

(c) The state with the largest positive residual appears to have a mean household income of about 67 thousand and to have about 48% graduating college, with a predicted percent graduating college (on the regression line) of about 37%. (This state is Massachusetts.)

(d) The state with the largest negative residual appears to have a mean household income of about 71 thousand dollars and to have about 25% graduating college, with a predicted percent graduating college of about 40%. (This state is Alaska.)

A.101 Answers will vary for all except part (a).

(a) The table of frequencies for the regions is given below.

Region	Frequency
1. Latin America	24
2. Western Nations	24
3. Middle East	16
4. Sub-Saharan Africa	33
5. South Asia	7
6. East Asia	12
7. Former Communist Countries	27
Total	143

Section 3.1 Solutions

3.1 This mean is a population parameter; notation is μ.

3.3 This proportion is a sample statistic; notation is $\hat{p}$.

3.5 This mean is a sample statistic; notation is $\bar{x}$.

3.7 This is a population parameter for a mean, so the correct notation is μ. We have $\mu = 59,388/148 = 401.3$ students as the average enrollment per charter school.

3.9 This is a sample statistic from a sample of size $n = 200$ for a correlation, so the correct notation is r. We have $r = 0.037$.

3.11 Since the data are for all regular players on the team, we need the population parameter for a correlation, so the correct notation is ρ. We use technology to see that $\rho = -0.147$.

3.13 We expect the sampling distribution to be centered at the value of the population mean, so we estimate that the population parameter is $\mu = 85$. The standard error is the standard deviation of the distribution of sample means. The middle of 95% of the distribution goes from about 45 to 125, about 40 on either side of $\mu = 85$. By the 95% rule, we estimate that $SE \approx 40/2 = 20$. (Answers may vary slightly.)

3.15 We expect the sampling distribution to be centered at the value of the population proportion, so we estimate that the population parameter is $p = 0.80$. The standard error is the standard deviation of the distribution of sample proportions. The middle of 95% of the distribution goes from about 0.74 to 0.86, about 0.06 on either side of $p = 0.80$. By the 95% rule, we estimate that $SE \approx 0.06/2 = 0.03$. (Answers may vary slightly.)

3.17 (a) We see in the sampling distribution that a sample mean of $\bar{x} = 70$ is not unusual for samples of this size, so this value is (*i*): reasonably likely to occur.

  (b) We see in the sampling distribution that a sample mean of $\bar{x} = 100$ is not unusual for samples of this size, so this value is (*i*): reasonably likely to occur.

  (c) We see in the sampling distribution that a sample mean of $\bar{x} = 140$ is rare for a sample of this size but similar sample means occurred several times in this sampling distribution. This value is (*ii*): unusual but might occur occasionally.

3.19 (a) We see in the sampling distribution that a sample proportion of $\hat{p} = 0.72$ is rare for a sample of this size but similar sample proportions occurred several times in this sampling distribution. This value is (*ii*): unusual but might occur occasionally.

  (b) We see in the sampling distribution that a sample proportion of $\hat{p} = 0.88$ is rare for a sample of this size but similar sample proportions occurred several times in this sampling distribution. This value is (*ii*): unusual but might occur occasionally.

  (c) We see in the sampling distribution that there are no sample proportions even close to $\hat{p} = 0.95$ so this sample proportion is (*iii*): extremely unlikely to ever occur using samples of this size.

3.21 (a) The parameter of interest is a population mean, so the notation is μ. We have $\mu =$ the mean number of apps downloaded by all smartphone users in the US who have downloaded at least one app.

(b) The quantity that gives the best estimate is the sample mean, which is denoted $\bar{x}$. Its value is $\bar{x} = 19.7$.

(c) We would have to ask all smartphone users in the US how many apps they have downloaded.

3.23 (a) We are estimating ρ, the correlation between pH and mercury levels of fish for all the lakes in Florida. The quantity that gives the best estimate is our sample correlation $r = -0.575$. We estimate that the correlation between pH levels and levels of mercury in fish in all Florida lakes is -0.575.

(b) We use an estimate because it would be very difficult and costly to find the exact population correlation. We would need to measure the pH level and the mercury in fish level for all the lakes in Florida, and there are over 7700 of them.

3.25 (a) The two distributions centered at the population average are probably unbiased, distributions A and D. The two distributions not centered at the population average ($\mu = 2.61$) are biased, dotplots B and C. The sampling for Distribution B gives an average too high, and has large households over-represented. The sampling for Distribution C gives an average too low and may have been done in an area with many people living alone.

(b) The larger the sample size the lower the variability, so distribution A goes with samples of size 100, and distribution D goes with samples of size 500.

3.27 The quantity we are trying to estimate is $\mu_m - \mu_o$ where μ_m represents the average grade for all fourth-grade students who study mixed problems and μ_o represents the average grade for all fourth-grade students who study problems one type at a time. The quantity that gives the best estimate is $\bar{x}_m - \bar{x}_o$, where $\bar{x}_m$ represents the average grade for the fourth-grade students in the sample who studied mixed problems and $\bar{x}_o$ represents the average grade for the fourth-grade students in the sample who studied problems one type at a time. The best estimate for the difference in the average grade based on study method is $\bar{x}_m - \bar{x}_o = 77 - 38 = 39$.

3.29 (a) We expect means of samples of size 30 to be much less spread out than values of budgets of individual movies. This leads us to conclude that Boxplot A represents the sampling distribution and Boxplot B represents the values in a single sample. We can also consider the shapes. Boxplot A appears to be symmetric and Boxplot B appears to be right skewed. Since we expect a sampling distribution to be symmetric and bell-shaped, Boxplot A is the sampling distribution and the skewed Boxplot B shows values in a single sample.

(b) Boxplot B shows the data from one sample of size 30. Each data value represents the budget, in millions of dollars, for one Hollywood movie made between 2007 and 2013. There are 30 values included in the sample. The budgets range from about 1 million to around 225 million for this sample. We see in the boxplot that the median is just over 40 million dollars. Since the data are right skewed, we expect the mean to be higher. We estimate the mean to be about 50 million or 55 million. This is the mean of a sample, so we have $\bar{x} \approx 53$ million dollars. (Answers may vary.)

(c) Boxplot A shows the data from a sampling distribution using samples of size 30. Each data value represents the *mean* of one of these samples. There are 1000 means included in the distribution. They range from about 30 to 90 million dollars. The center of the distribution is a good estimate of the population parameter, and the center appears to be about $\mu \approx 56$ million dollars, where μ represents the mean budget, in millions of dollars, for *all* movies coming out of Hollywood between 2007 and 2013. (Answers may vary.)

3.31 (a) Answers will vary. The following table gives one possible set of randomly selected Enrollment values. The mean for this sample is $\overline{x} = 53.0$.

University	Department	Enrollment
Case Western Reserve University	Statistics	11
University of South Carolina	Biostatistics	45
Harvard University	Statistics	67
University of California - Riverside	Statistics	54
Medical University of South Carolina	Biostatistics	46
University of Nebraska	Statistics	44
New York University	Statistics	6
Columbia University	Statistics	196
University of Iowa	Biostatistics	35
Baylor University	Statistics	26

(b) Answers will vary. The following table gives another possible set of randomly selected Enrollment values. The sample mean for this sample is $\overline{x} = 61.5$.

University	Department	Enrollment
University of Wisconsin	Statistics	116
Cornell University	Statistics	78
Yale University	Statistics	36
Iowa State University	Statistics	145
Boston University	Biostatistics	39
University of Nebraska	Statistics	44
University of Minnesota	Biostatistics	48
University of California - Los Angeles	Biostatistics	60
University of California - Davis	Statistics	34
Virginia Commonwealth University	Statistics	15

(c) The population mean number of enrollment for all 82 statistics graduate programs $\mu = 53.54$. Most sample means found in parts (a) and (b) will be somewhat close to this but not exactly the same.

(d) The distribution will be roughly symmetric and centered at 53.54.

3.33 Answers will vary, but a typical distribution is shown below. The sampling distribution is symmetric and bell-shaped, and is centered approximately at the population mean of $\mu = 53.54$. The standard error for this set of simulated means is about 10.9.

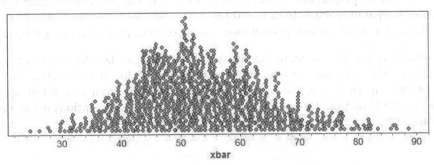

3.35 (a) The mean and standard deviation from the population of all Hollywood movies made between 2007 and 2013 are parameters so the correct notation is μ and σ respectively. Based on the *Budget* variable in **HollywoodMovies**, we find $\mu = 56.12$ million dollars and $\sigma = 53.76$ million dollars. Note that several movies in the dataset are missing values for the *Budget* variable, so we will consider the population to be all movies that have *Budget* values.

(b) Using technology we produce a sampling distribution (shown below) with means for 1000 samples of size $n = 20$ taken from the movie budgets in the full dataset. We see that the distribution is symmetric, bell-shaped, and centered at the population mean of \$56.1 million. The standard deviation of these 1000 sample means is 11.9, so we estimate the standard error for Hollywood movie budgets based on samples of 20 movies to be $SE \approx \$12$ million.

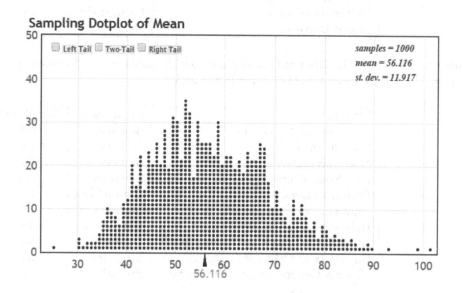

3.37 (a) This is a population proportion so the correct notation is p. We have $p = 47/303 = 0.155$.

(b) We expect it to be symmetric and bell-shaped and centered at the population proportion of 0.155.

3.39 (a) The standard error is the standard deviation of the sampling distribution (given in the upper right corner of the sampling distribution box of *StatKey*) and is likely to be about 0.11. Answers will vary, but the sample proportions should go from 0 to about 0.6 (as in the dotplot below). In that case, the farthest sample proportion from $p = 0.155$ is $\hat{p} \approx 0.6$, and it is $0.6 - 0.155 = 0.445$ off from the correct population value. In other simulations the maximum proportion might be as high as 0.7.

(b) The standard error is the standard deviation of the sampling distribution and is likely to be about 0.08. Answers will vary, but the sample proportions should go from 0 to about 0.5 (as shown in the dotplot below). In that case, the farthest sample proportion from $p = 0.155$ is $\hat{p} \approx 0.5$, and it is $0.5 - 0.155 = 0.345$ off from the correct population value. Some simulations might produce even larger discrepancies.

(c) The standard error is the standard deviation of the sampling distribution and is likely to be about 0.05. Answers will vary, but the sample proportions should go from near 0 to about 0.3 (0.02 to 0.32 in the dotplot below). In that case, the farthest sample proportion from $p = 0.155$ is $\hat{p} \approx 0.32$, and it is $0.32 - 0.155 = 0.165$ off from the correct population value. Some simulations might have even larger discrepancies.

(d) Accuracy improves as the sample size increases. The standard error gets smaller, the range of values gets smaller, and values tend to be closer to the population value of $p = 0.155$.

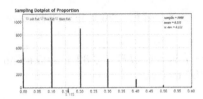 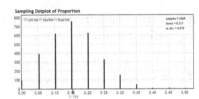 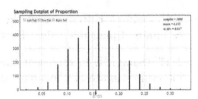

Section 3.2 Solutions

3.41   (a) We are estimating a population proportion, so the notation is p.

(b) The best estimate of the population proportion is the sample proportion $\hat{p}$.

3.43   (a) We are estimating a difference in means, so the notation is $\mu_1 - \mu_2$.

(b) The best estimate of the difference in population means is the difference in sample means $\bar{x}_1 - \bar{x}_2$.

3.45 Using ME to represent the margin of error, an interval estimate for μ is $\bar{x} \pm ME = 25 \pm 3$ so an interval estimate of plausible values for the population mean μ is 22 to 28.

3.47 Using ME to represent the margin of error, an interval estimate for ρ is $r \pm ME = 0.62 \pm 0.05$ so an interval estimate of plausible values for the population correlation ρ is 0.57 to 0.67.

3.49   (a) Yes, plausible values of μ are values in the interval.

(b) Yes, plausible values of μ are values in the interval.

(c) No. Since 105.3 is not in the interval estimate, it is a *possible* value of μ but is not a very plausible one.

3.51 The 95% confidence interval estimate is $\hat{p} \pm 2 \cdot SE = 0.32 \pm 2(0.04) = 0.32 \pm 0.08$, so the interval is 0.24 to 0.40. We are 95% confident that the true value of the population proportion p is between 0.24 and 0.40.

3.53 The 95% confidence interval estimate is $r \pm 2 \cdot SE = 0.34 \pm 2(0.02) = 0.34 \pm 0.04$, so the interval is 0.30 to 0.38. We are 95% confident that the true value of the population correlation ρ is between 0.30 and 0.38.

3.55 The 95% confidence interval estimate is $(\bar{x}_1 - \bar{x}_2) \pm$ margin of error $= 3.0 \pm 1.2$, so the interval is 1.8 to 4.2. We are 95% confident that the true difference in the population means $\mu_1 - \mu_2$ is between 1.8 and 4.2 (which means we believe that the mean of population 1 is between 1.8 and 4.2 units larger than the mean of population 2.)

3.57   (a) We are estimating a difference in means, so we are estimating $\mu_A - \mu_T$, where μ_A represents the mean fear response for adults and μ_T represents the mean fear response for teenagers.

(b) The best estimate is given by the difference in sample means $\bar{x}_A - \bar{x}_T$. Its value is $\bar{x}_A - \bar{x}_T = 0.225 - 0.059 = 0.166$.

(c) The 95% confidence interval is given by

$$
\begin{aligned}
\text{Statistic} &\pm 2 \cdot SE \\
(\bar{x}_A - \bar{x}_T) &\pm 2 \cdot SE \\
0.166 &\pm 2(0.091) \\
0.166 &\pm 0.182 \\
-0.016 \quad &\text{to} \quad 0.348.
\end{aligned}
$$

The 95% confidence interval for the difference in mean fear response is -0.016 to 0.348.

(d) This is an observational study since the explanatory variable (age) was not manipulated.

3.59   (a) We are estimating a difference in proportions, so the notation is $p_m - p_f$, where we define $p_m =$ the proportion of male college students who choose pain over solitude and $p_f =$ the proportion of female college students who choose pain over solitude.

(b) The best estimate is given by the difference in sample proportions $\hat{p}_m - \hat{p}_f$. We have $\hat{p}_m - \hat{p}_f = 12/18 - 6/24 = 0.667 - 0.25 = 0.417$.

(c) The 95% confidence interval is given by

$$
\begin{aligned}
\text{Statistic} &\pm 2 \cdot SE \\
(\hat{p}_m - \hat{p}_f) &\pm 2 \cdot SE \\
0.417 &\pm 2(0.154) \\
0.417 &\pm 0.308 \\
0.109 \quad &\text{to} \quad 0.725.
\end{aligned}
$$

The 95% confidence interval for the difference in proportion of college students who would choose pain over solitude, between males and females, is 0.109 to 0.725.

(d) "No difference" corresponds to a difference of zero. Since zero is not in the confidence interval, this is not a plausible value. Since all plausible values are positive, males appear to be more likely than females to choose pain over solitude.

3.61 (a) The information is from a sample, so it is a statistic. It is a proportion, so the correct notation is $\hat{p} = 0.30$.

(b) The parameter we are estimating is the proportion, p, of *all* young people in the US who have been arrested by the age of 23. Using the information in the sample, we estimate that $p \approx 0.30$.

(c) If the margin of error is 0.01, the interval estimate is 0.30 ± 0.01 which gives 0.29 to 0.31. Plausible values for the proportion p range from 0.29 to 0.31.

(d) Since the plausible values for the true proportion are those between 0.29 and 0.31, it is very unlikely that the actual proportion is less than 0.25.

3.63 We are 95% confident that the proportion of all adults in the US who think a car is a necessity is between 0.83 and 0.89.

3.65 We are estimating p, the proportion of all US adults who agree with the statement that each person has one true love. The best point estimate is $\hat{p} = 735/2625 = 0.28$. We find the confidence interval using:

$$
\begin{aligned}
\hat{p} &\pm 2 \cdot SE \\
0.28 &\pm 2(0.009) \\
0.28 &\pm 0.018 \\
0.262 \quad &\text{to} \quad 0.298.
\end{aligned}
$$

The margin of error for our estimate is 0.018 or 1.8%. We are 95% sure that the proportion of all US adults who agree with the statement on one true love is between 0.262 and 0.298.

3.67 (a) We are 95% confident that the mean response time for game players minus the mean response time for non-players is between -1.8 to -1.2. In other words, mean response time for game players is less than the mean response time for non-players by between 1.8 and 1.2 seconds.

(b) It is not likely that they are basically the same, since the option of the difference in means being zero is not in the interval. The game players are faster, and we can tell this because the confidence interval for $\mu_g - \mu_{ng}$ has only negative values so the mean time is smaller for the game players.

(c) We are 95% confident that the mean accuracy score for game players minus the mean accuracy score for non-players is between -4.2 to 5.8.

(d) It is likely that they are basically the same, since the option of the difference in means being zero is in the interval. There is little discernible difference in accuracy between game players and non-game players.

3.69 (a) Using the margin of error, we see that the likely proportion voting for Candidate A ranges from 49% to 59%. Since this interval includes some proportions below 50% as plausible values for the election proportion, we cannot be very confident in the outcome.

(b) Using the margin of error, we see that the likely proportion voting for Candidate A ranges from 51% to 53%. Since all values in this interval are over 50%, we can be relatively confident that Candidate A will win.

(c) Using the margin of error, we see that the likely proportion voting for Candidate A ranges from 51% to 55%. Since all values in this range are over 50%, we can be relatively confident that Candidate A will win.

(d) Using the margin of error, we see that the likely proportion voting for Candidate A ranges from 48% to 68%. Since this interval includes some proportions below 50% as plausible vaues for the election proportion, we cannot be very confident in the outcome.

3.71 Let μ represent the mean time for a golden shiner fish to find the yellow mark. A 95% confidence interval is given by

$$
\begin{aligned}
\bar{x} \;\pm\; & 2 \cdot SE \\
51 \;\pm\; & 2(2.4) \\
51 \;\pm\; & 4.8 \\
46.2 \quad \text{to} \quad & 55.8.
\end{aligned}
$$

A 95% confidence interval for the mean time for fish to find the mark is between 46.2 and 55.8 seconds. We are 95% sure that the *mean* time it would take fish to find the target for *all* fish of this breed is between 46.2 seconds and 55.8 seconds. In other words, the plausible values for the population mean μ are those values between 46.2 and 55.8. Therefore, 60 is not a plausible value for the mean time for all fish, but 55 is.

3.73 We are estimating the difference in population proportions $p_1 - p_2$ where p_1 is the proportion of times a school of fish will pick the majority option if there is an opinionated minority, a less passionate majority, and also some additional members with no preference and p_2 is the proportion of times a school of fish will pick the majority option if there is an opinionated minority and a less passionate majority and no other fish in the group, as described above in *Fish Democracies*. (We could also have defined the proportions in the other order.) The best point estimate is $\hat{p}_1 - \hat{p}_2 = 0.61 - 0.17 = 0.44$. We find a 95% confidence interval as follows:

$$
\begin{aligned}
(\hat{p}_1 - \hat{p}_2) \;\pm\; & 2 \cdot SE \\
(0.61 - 0.17) \;\pm\; & 2(0.14) \\
0.44 \;\pm\; & 0.28 \\
0.16 \quad \text{to} \quad & 0.72.
\end{aligned}
$$

We are 95% sure that the proportion of schools of fish picking the majority option is 0.16 to 0.72 higher if fish with no preference are added to the group. If adding the indifferent fish had no effect, then the population proportions with and without the indifferent fish would be the same, which means the difference in proportions would be zero. Since zero is not a plausible value for the difference in proportions, it is very unlikely that adding indifferent fish has no effect. The indifferent fish are helping the majority carry the day.

Section 3.3 Solutions

3.75 (a) No. The value 12 is not in the original.

(b) No. A bootstrap sample has the same sample size as the original sample.

(c) Yes.

(d) No. A bootstrap sample has the same sample size as the original sample.

(e) Yes.

3.77 The distribution appears to be centered near 0.7 so the point estimate is about 0.7. Using the 95% rule, we estimate that the standard error is about 0.1 (since about 95% of the values appear to be within 0.2 of the center). Thus our interval estimate is

$$
\begin{array}{ccc}
\text{Statistic} & \pm & 2 \cdot SE \\
0.7 & \pm & 2(0.1) \\
0.7 & \pm & 0.2 \\
0.5 & \text{to} & 0.9.
\end{array}
$$

The parameter being estimated is a proportion p, and the interval 0.5 to 0.9 gives plausible values for the population proportion p. Answers may vary.

3.79 The distribution appears to be centered near 0.4 so the point estimate is about 0.4. Using the 95% rule, we estimate that the standard error is about 0.05 (since about 95% of the values appear to be within 0.1 of the center). Thus our interval estimate is

$$
\begin{array}{ccc}
\text{Statistic} & \pm & 2 \cdot SE \\
0.4 & \pm & 2(0.05) \\
0.4 & \pm & 0.1 \\
0.3 & \text{to} & 0.5.
\end{array}
$$

The parameter being estimated is a correlation ρ, and the interval 0.3 to 0.5 gives plausible values for the population correlation ρ. Answers may vary.

3.81 The statistic for the sample is $\hat{p} = 35/100 = 0.35$. Using technology, the standard deviation of the sample proportions for 1000 bootstrap samples is about 0.048 (answers may vary slightly), so we estimate the standard error is SE≈ 0.048. Thus our interval estimate is

$$
\begin{array}{ccc}
\text{Statistic} & \pm & 2 \cdot SE \\
0.35 & \pm & 2(0.048) \\
0.35 & \pm & 0.096 \\
0.254 & \text{to} & 0.446.
\end{array}
$$

Plausible values of the population proportion range from 0.254 to 0.446.

3.83 The statistic for the sample is $\hat{p} = 112/400 = 0.28$. Using technology, the standard deviation of the sample proportions for 1000 bootstrap samples is about 0.022 (answers may vary slightly), so we estimate the standard error is SE≈ 0.022. Thus our interval estimate is

$$
\begin{array}{ccc}
\text{Statistic} & \pm & 2 \cdot SE \\
0.28 & \pm & 2(0.022) \\
0.28 & \pm & 0.044 \\
0.236 & \text{to} & 0.324.
\end{array}
$$

Plausible values of the population proportion range from 0.236 to 0.324.

3.85 (a) The best point estimate is the sample proportion, $\hat{p} = 26/174 = 0.149$.

(b) We can estimate the standard error using the 95% rule, or we can find the standard deviation of the bootstrap statistics in the upper right of the figure. We see that the standard error is about 0.028. Answers will vary slightly with other simulations.

(c) We have

$$
\begin{array}{ccc}
\hat{p} & \pm & 2 \cdot SE \\
0.149 & \pm & 2(0.028) \\
0.149 & \pm & 0.056 \\
0.093 & \text{to} & 0.205.
\end{array}
$$

We are 95% confident that the percent of all snails of this kind that will live after being eaten by a bird is between 9.3% and 20.5%.

(d) Yes, 20% is within the range of plausible values in the 95% confidence interval.

3.87 (a) The original data is shown in Dotplot I which has only 25 dots and a lot of variability. The bootstrap distribution for many sample means is shown in Dotplot II.

(b) The sample mean hippocampal volume appears to be about $\bar{x} = 7600$. This can be seen in either the original data dotplot or (more easily) the bootstrap distribution.

(c) We need to use the bootstrap distribution (Dotplot II) to estimate the standard error. Using the 95% rule, the standard error appears to be approximately $SE \approx 200$.

(d) The standard deviation of the original data (in Dotplot I) appears to be larger than the standard error, as the data in the original sample has more variability than the simulated sample means in the bootstrap distribution.

(e) The rough 95% confidence interval is given by

$$
\begin{array}{ccc}
\text{Statistic} & \pm & 2 \cdot SE \\
\bar{x} & \pm & 2 \cdot SE \\
7600 & \pm & 2(200) \\
7600 & \pm & 400 \\
7200 & \text{to} & 8000.
\end{array}
$$

We are 95% confident that mean hippocampal volume, in μL, for non-football playing people is between 7200 and 8000.

3.89 The sample proportion of females showing compassion is $\hat{p}_F = 6/6 = 1.0$. The sample proportion of males showing compassion is $\hat{p}_M = 17/24 = 0.708$. The best point estimate for the difference in proportions $p_F - p_M$ is $\hat{p}_F - \hat{p}_M = 1.0 - 0.708 = 0.292$. Using *StatKey* to create a bootstrap distribution for a difference in proportions using this sample data, we see a standard error of 0.094.

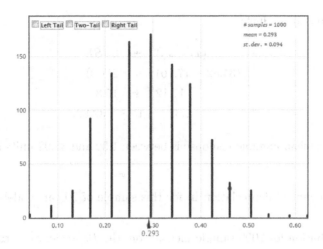

We have

$$
\begin{aligned}
(\hat{p}_F - \hat{p}_M) &\pm 2 \cdot SE \\
(1.0 - 0.708) &\pm 2(0.094) \\
0.292 &\pm 0.188 \\
0.104 \quad &\text{to} \quad 0.480.
\end{aligned}
$$

Based on this interval the percentage of female rats likely to show compassion is between 10.4% and 48% higher than the percentage of male rats likely to show compassion. Since zero is not in the interval estimate, it is not very plausible that male and female rats are equally compassionate.

3.91 Using *StatKey* or other technology, we create a bootstrap distribution to estimate the difference in means $\mu_t - \mu_c$ where μ_t represents the mean immune response for tea drinkers and μ_c represents the mean immune response for coffee drinkers. In the original sample the means are $\overline{x}_t = 34.82$ and $\overline{x}_c = 17.70$, respectively, so the point estimate for the difference is $\overline{x}_t - \overline{x}_c = 34.82 - 17.70 = 17.12$. We see from the bootstrap distribution that the standard error for the differences in bootstrap means is about $SE = 7.9$. This will vary for other sets of bootstrap differences.

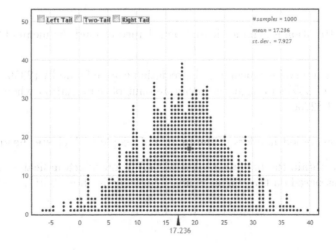

For a 95% confidence interval, we have

$$
\begin{aligned}
(\bar{x}_t - \bar{x}_c) \quad &\pm \quad 2 \cdot SE \\
(34.82 - 17.70) \quad &\pm \quad 2(7.9) \\
17.12 \quad &\pm \quad 15.8 \\
1.32 \quad &\text{to} \quad 32.92.
\end{aligned}
$$

We are 95% sure that the mean immune response is between 1.32 and 32.92 units higher in tea drinkers than it is in coffee drinkers.

3.93 (a) The mean amount of depreciation in for this sample of 20 car models is $2356 with a standard deviation of $858.

(b) A bootstrap distribution for 2000 sample means from the *Depreciation* variable is shown below.

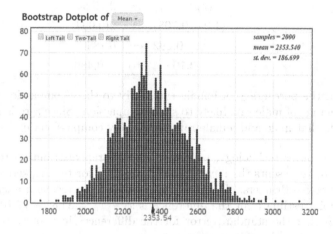

The bootstrap distribution is symmetric and bell shaped around the mean of 2354, and has a standard deviation of 187.

(c) The 95% confidence interval for mean car depreciation can be found by $(2356 - 2(187), 2356 + 2(187)) = (1982, 2730)$. We are 95% sure that the mean amount of depreciation when new cars leave the lot is between $1982 and $2730.

3.95 (a) The proportion of left-handers in the sample of people with cluster headaches is $\hat{p} = \frac{24}{273} = 0.088$.

(b) Using StatKey, we obtain the bootstrap distribution below which indicates the standard error for the sample proportions based on the cluster headache sample is about 0.017.

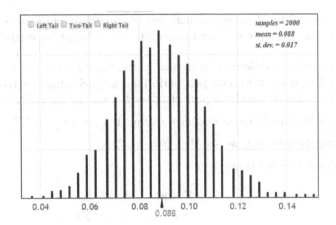

The confidence interval for the proportion is

$$0.088 \pm 2 \cdot 0.017 = 0.088 \pm 0.034 = 0.054 \text{ to } 0.122$$

We are 95% confident that the proportion of cluster headache sufferers who are left handed is between 0.054 and 0.122.

(c) The proportion of left-handers in the sample of people with migraine headaches is $\hat{p} = \frac{42}{477} = 0.088$.

(d) Using StatKey, we obtain the bootstrap distribution below which indicates the standard error for the sample proportions based on the migraine headache sample is about 0.013.

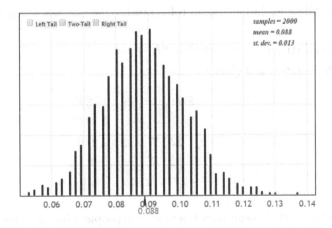

The confidence interval for the proportion is

$$0.088 \pm 2 \cdot 0.013 = 0.088 \pm 0.026 = 0.062 \text{ to } 0.114$$

We are 95% confident that the proportion of migraine headache sufferers who are left handed is between 0.062 and 0.114.

(e) The confidence interval for the migraine sufferers (d) is slightly more narrow. This is expected, because the sample proportions are similar but the sample size is larger for migraine sufferers.

3.97 (a) We are estimating a difference in means, so the notation is $\mu_S - \mu_N$, where μ_S represents mean closeness rating for people who have just performed a synchronized activity and μ_N represents mean closeness rating for people who have performed such a non-synchronized activity.

(b) We see that the difference in mean closeness ratings is $\overline{x}_S - \overline{x}_N = 5.275 - 4.810 = 0.465$.

(c) Using a bootstrap distribution, we see that $SE \approx 0.228$.

(d) The 95% confidence interval is given by

$$
\begin{array}{rcl}
\text{Statistic} & \pm & 2 \cdot SE \\
(\overline{x}_S - \overline{x}_N) & \pm & 2 \cdot SE \\
0.465 & \pm & 2(0.228) \\
0.465 & \pm & 0.456 \\
0.009 & \text{to} & 0.921.
\end{array}
$$

(e) We are 95% confident that people who have just done a synchronized activity give a mean closeness rating between 0.009 and 0.921 higher than people who have just done a non-synchronized activity.

3.99 (a) We are estimating a difference in means, so the notation is $\mu_H - \mu_L$, where μ_H represents mean pain tolerance for people who have just performed a high exertion activity and μ_L represents mean pain tolerance for those who have just performed a low exertion activity.

(b) We see that the difference in mean pain tolerance for the samples is $\overline{x}_H - \overline{x}_L = 224.130 - 211.413 = 12.717$.

(c) Using a bootstrap distribution, we see that $SE \approx 9.0$.

(d) The 95% confidence interval is given by

$$
\begin{array}{rcl}
\text{Statistic} & \pm & 2 \cdot SE \\
(\overline{x}_S - \overline{x}_N) & \pm & 2 \cdot SE \\
12.717 & \pm & 2(9.0) \\
12.717 & \pm & 18.0 \\
-5.283 & \text{to} & 30.717.
\end{array}
$$

(e) We are 95% confident that mean pain tolerance for people who have done a high exertion activity is between 5.283 mmHg below and 30.717 mmHg above the mean pain tolerance for those who have done a low exertion activity.

3.101 (a) We are estimating a difference in proportions, so the notation is $p_F - p_M$, where p_F represents the proportion of females who would allow the pressure to reach its maximum level and p_M represents the proportion of males who would allow the pressure to reach its maximum level.

(b) We have $\hat{p}_F = 33/165 = 0.200$ and $\hat{p}_M = 42/99 = 0.424$, so the difference in sample proportions is $\hat{p}_F - \hat{p}_M = 0.200 - 0.424 = -0.224$.

(c) Using a bootstrap distribution, we see that $SE \approx 0.059$.

(d) The 95% confidence interval is given by

$$
\begin{array}{ccc}
\text{Statistic} & \pm & 2 \cdot SE \\
(\hat{p}_F - \hat{p}_M) & \pm & 2 \cdot SE \\
-0.224 & \pm & 2(0.059) \\
-0.224 & \pm & 0.118 \\
-0.342 & \text{to} & -0.106.
\end{array}
$$

(e) We are 95% confident that the proportion of females able to "go to the max" is between 0.342 and 0.106 lower than the proportion of males able to do this. Females appear to have lower pain tolerance.

3.103 The standard deviation for the sample of penalty minutes for n=24 players is $s = 27, 3$ minutes. For one set of 3000 bootstrap sample standard deviations (shown below), the estimated standard error is $SE = 5.2$.

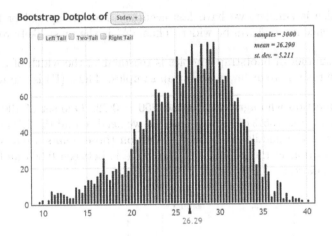

Based on this the interval estimate is

$$
\begin{array}{ccc}
s & \pm & 2 \cdot SE \\
27.3 & \pm & 2 \cdot 5.2 \\
27.3 & \pm & 10.4 \\
16.9 & \text{to} & 37.7.
\end{array}
$$

We estimate that the standard deviation in penalty minutes for all NHL players is somewhere between 16.9 and 37.7 minutes.

Section 3.4 Solutions

3.105 (a) We keep the middle 95% of values by chopping off 2.5% from each tail. Since 2.5% of 1000 is 25, we eliminate the 25 highest and the 25 lowest values to create the 95% confidence interval.

(b) We keep the middle 90% of values by chopping off 5% from each tail. Since 5% of 1000 is 50, we eliminate the 50 highest and the 50 lowest values to create the 90% confidence interval.

(c) We keep the middle 98% of values by chopping off 1% from each tail. Since 1% of 1000 is 10, we eliminate the 10 highest and the 10 lowest values to create the 98% confidence interval.

(d) We keep the middle 99% of values by chopping off 0.5% from each tail. Since 0.5% of 1000 is 5, we eliminate the 5 highest and the 5 lowest values to create the 99% confidence interval.

3.107 To find a 90% confidence interval, we go less far out on either side than for a 95% confidence interval, so (C) is the most likely result.

3.109 If the sample size is smaller, we have less accuracy and the spread of the bootstrap distribution increases, so the confidence interval will be wider. Thus, (A) is the most likely result.

3.111 As long as the number of bootstrap samples is reasonable, the width of the confidence interval does not change much as we take more or fewer bootstrap samples. Thus, (B) is the most likely result.

3.113 The sample proportion who agree is $\hat{p} = 180/250 = 0.72$. One set of 1000 bootstrap proportions is shown in the figure below. For a 95% confidence interval we need to find the 2.5%-tile and 97.5%-tile, leaving 95% of the distribution in the middle. For this distribution those points are at 0.664 and 0.776, so we are 95% sure that the proportion in the population who agree is between 0.664 and 0.776. Answers will vary slightly for different simulations.

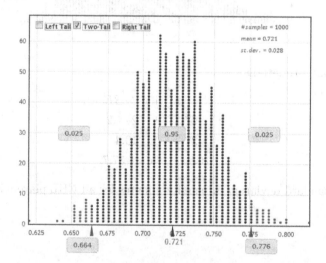

3.115 The sample proportion who agree is $\hat{p} = 382/1000 = 0.382$. One set of 1000 bootstrap proportions is shown in the figure below. For a 99% confidence interval we need to find the 0.5%-tile and 99.5%-tile, leaving 99% of the distribution in the middle. For this distribution those points are at 0.343 and 0.423, so we are 99% sure that the proportion in the population who agree is between 0.343 and 0.423. Answers will vary slightly for different simulations.

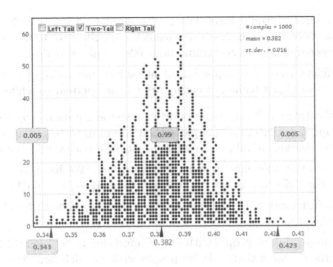

3.117 The 98% confidence interval uses the 1%-tile and 99%-tile from the bootstrap means. We are 98% sure that the mean number of penalty minutes for NHL players in a season is between 22.8 and 47.7 minutes.

3.119 Using *StatKey* or other technology, we produce a bootstrap distribution such as the figure shown below. For a 99% confidence interval, we find the 0.5%-tile and 99.5%-tile points in this distribution to be 0.467 and 0.493. We are 99% confident that the percent of all Europeans (from these nine countries) who can identify arm or shoulder pain as a symptom of a heart attack is between 46.7% and 49.3%. Since every value in this interval is below 50%, we can be 99% confident that the proportion is less than half.

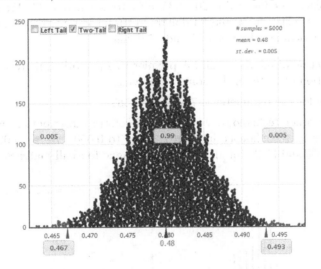

3.121 (a) For a 99% confidence interval, we want 99% in the middle, which means leaving 1% total in the tails, so 0.5% in each tail. Therefore, we are looking for the 0.5^{th} and 99.5^{th} percentiles, which are 15.8 and 32.4. Therefore, our 99% confidence interval is (15.8, 32.4) μg/g crt.

   (b) We are 99% confident that, on average, concentration of 3-PBA is between 15.8 and 32.4 μg/g crt higher while not eating organic as opposed to while eating organic.

3.123 (a) The sample difference in proportions is $\hat{p}_1 - \hat{p}_2 = 111/240 - 24/240 = 0.46 - 0.10 = 0.36$.

(b) A 98% confidence interval would leave 98% in the middle, so 1% = 0.01 in each tail. Because there are 1000 simulated bootstrap samples, this means we should leave $0.01 \times 1000 = 10$ dots in each tail, resulting in a 98% confidence interval from about 0.28 to about 0.45.

(c) We are 98% confident that the proportion of measurements yielding positive pesticide detection is between 0.28 and 0.45 higher while not eating organic as opposed to while eating organic.

3.125 (a) We are estimating a difference in proportions, so the notation is $p_T - p_N$, where p_T is proportion of people with staph infections of all people with triclosan in their system and p_N is the proportion of people with staph infections of all people who do not have triclosan in their systems.

(b) The best estimate is the difference in sample proportions. We have $\hat{p}_T = 24/37 = 0.649$ and $\hat{p}_N = 15/53 = 0.283$, so the difference in sample proportions is $\hat{p}_T - \hat{p}_N = 0.649 - 0.283 = 0.366$.

(c) Using a bootstrap distribution, we see that a 99% confidence interval for the difference in proportions is approximately 0.100 to 0.614.

(d) We see that the proportion of people with staph infections is between 0.10 to 0.61 higher for people with triclosan in their system than it is for people without triclosan. Yes, since all of these plausible values for the difference in proportions are positive, we can conclude that people with triclosan in their system are more likely to have staph infections. We cannot conclude causation, however, since this is not an experiment.

3.127 (a) A 99% confidence interval is wider than a 90% confidence interval, so the 90% interval is A (3.55 to 4.15) and the 99% interval is B (3.35 to 4.35).

(b) We multiply the lower and upper bounds for the average tip by 20 to get the average daily tip revenue (assuming 20 tables per day). With 90% confidence, the interval is $20 \cdot 3.55 = 71$ to $20 \cdot 4.15 = 83$. With 99% confidence, the interval is $20 \cdot 3.35 = 67$ to $20 \cdot 4.35 = 87$. We are 90% confident that this waitress will average between 71 and 83 dollars in tip income per day, and we are 99% confident that her mean daily tip income is between 67 and 87 dollars.

3.129 (a) The population of interest is all FA premier league football matches. The specific parameter of interest is proportion of matches the home team wins.

(b) Our best estimate for the parameter is $70/120 = 0.583$.

(c) Using *StatKey* or other technology, we create a bootstrap distribution as shown below. Taking 5% from each tail, the 90% confidence interval is 0.508 to 0.650. We are 90% sure that the home team wins between 50.8% and 65.0% of all FA premier league football matches.

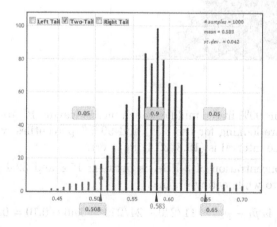 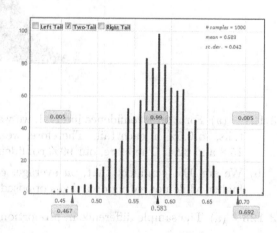

(d) Using the same bootstrap distribution we see that a 99% confidence interval goes from 0.467 to 0.692. We are 99% sure that the home team wins between 46.7% and 69.2% of all FA premier league football matches.

(e) If the population parameter is 0.50 or less, then no home field advantage is present. With the 90% confidence interval we are 90% confident the population parameter is between 0.508 and 0.650. Since this interval does not contain 0.50, we are 90% confident that there is a home field advantage. However the 99% confidence interval does contain 0.50, so we are not 99% confident that there is a home field advantage.

3.131 The difference in sample means is $\overline{x}_I - \overline{x}_S = 37.29 - 50.92 = -13.63$. A bootstrap distribution for differences in means is shown below.

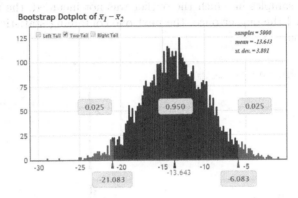

Using the bootstrap standard error ($SE = 3.80$) we get a 95% confidence interval with

$$(\overline{x}_I - \overline{x}_S) \pm 2 \cdot SE = -13.63 \pm 2 \cdot 3.80 = -13.63 \pm 7.60 = (-21.23, -6.03)$$

Thus we conclude that we are 95% sure that the mean cost when paying individually is somewhere between 21.23 and 6.03 shekels less than when splitting the bill.

Using percentiles from this bootstrap distribution the interval would go from -21.08 shekels to -6.08 shekels.

3.133 (a) The parameter of interest is ρ, the correlation between weight gain during a month of overeating and inactivity and weight gain over the next 2.5 years, for those adults who spend one month (possibly during December) overeating and being sedentary. The best point estimate for this parameter is $r = 0.21$.

(b) To create the bootstrap sample, we sample from the original sample with replacement. In this case, we randomly select one of the 18 ordered pairs, write down the values, and return them to the pile. Then we randomly select one of the 18 ordered pairs (possibly the same one), and write down those values as our second pair. We do this until we have 18 ordered pairs, and that dataset is our bootstrap sample.

(c) For each bootstrap sample, we record the correlation between the one month and 2.5 year weight gains of the 18 ordered pairs.

(d) We find the standard error by finding the standard deviation of the 1000 bootstrap correlations.

(e) The interval estimate is $r \pm 2 \cdot SE = 0.21 \pm 2(0.14) = 0.21 \pm 0.28$, so a 95% confidence interval for the population correlation ρ is -0.07 to 0.49.

(f) There is a reasonable possibility that there is no correlation at all between the amount of weight gained during the one month intervention and how much weight is gained over the long-term. We know that this is a reasonable possibility because 0 is inside the interval estimate so $\rho = 0$ is included as one of the plausible values of the population correlation.

(g) A 90% confidence interval needs to only include the middle 90% of data values in a bootstrap distribution, so it will be narrower than a 95% confidence interval.

3.135 The bootstrap distribution for the standard deviations (shown below) has at least four completely separate clusters of dots. It is not at all symmetric and bell-shaped so it would *not* be appropriate to use this bootstrap distribution to find a confidence interval for the standard deviation. The clusters of dots represent the number of times the outlier is included in the bootstrap sample (with the cluster on the left containing statistics from samples in which the outlier was not included, the next one containing statistics from samples that included the outlier once, the next one containing statistics from samples that included the outlier twice, and so on.)

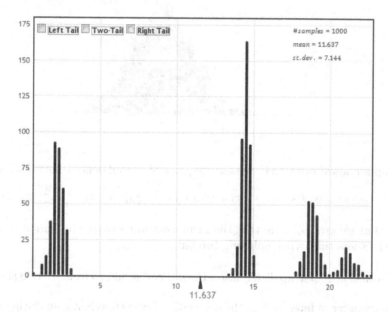

Section 4.1 Solutions

4.1 (a) We see that Sample A has the largest mean (around 30) and only one data point below 25, so it provides the most evidence for the claim that the mean placement exam score is greater than 25.

(b) Sample C has a mean that is clearly below 25, so it provides no evidence for the claim.

4.3 (a) Sample A shows a negative association and a stronger association than Sample D, so Sample A provides the most evidence for the claim that the correlation between exam grades and time spent playing video games is negative.

(b) In both samples B and C the the association in the scatterplots is positive, so both give no evidence for a negative correlation.

4.5 The hypotheses are:

$$H_0 : \quad \mu_A = \mu_B$$
$$H_a : \quad \mu_A \neq \mu_B$$

4.7 The hypotheses are:

$$H_0 : \quad \mu = 50$$
$$H_a : \quad \mu < 50$$

4.9 We define p_m to be the proportion of males who smoke and p_f to be the proportion of females who smoke. The hypotheses are:

$$H_0 : \quad p_m = p_f$$
$$H_a : \quad p_m > p_f$$

4.11 We define p to be the proportion of a population who watch the Home Shopping Network. The hypotheses are:

$$H_0 : \quad p = 0.20$$
$$H_a : \quad p < 0.20$$

4.13 We define μ_f to be mean study time for first year students and μ_u to be mean study time for upperclass students. The hypotheses are:

$$H_0 : \quad \mu_f = \mu_u$$
$$H_a : \quad \mu_f \neq \mu_u$$

4.15 (a) These hypotheses are valid.

(b) These hypotheses are not valid, since statistical hypotheses are statements about a population parameter (p), not a sample statistic ($\hat{p}$).

(c) These hypotheses are not valid, since the equality should be in H_0.

(d) These hypotheses are not valid, since a proportion, p, is always between 0 and 1 and can never be 25.

4.17 (a) We define μ_b to be the mean number of mosquitoes attracted after drinking beer and μ_w to be the mean number of mosquitoes attracted after drinking water. The hypotheses are:

$$H_0: \quad \mu_b = \mu_w$$
$$H_a: \quad \mu_b > \mu_w$$

(b) The sample mean number of mosquitoes attracted per participant before consumption for the beer group is $434/25 = 17.36$ and is $337/18 = 18.72$ for the water group. These sample means are slightly different, but the small difference could be attributed to random chance.

(c) The sample mean number of mosquitoes attracted per participant after consumption is $590/25 = 23.60$ for the beer group and is $345/18 = 19.17$ for the water group. This difference is larger than the difference in means before consumption. It is less likely to be due just to random chance.

(d) The mean number of mosquitoes attracted when drinking beer is higher than when drinking water.

(e) Since this was an experiment we can conclude causation and, if there is evidence for the alternative hypothesis, we have evidence that beer consumption increases mosquito attraction.

4.19 (a) We define μ_e to be mean BMP level in the brains of exercising mice and μ_s to be mean BMP level in the brains of sedentary mice. The hypotheses are:

$$H_0: \quad \mu_e = \mu_s$$
$$H_a: \quad \mu_e < \mu_s$$

(b) We define μ_e to be mean noggin level in the brains of exercising mice and μ_s to be mean noggin level in the brains of sedentary mice. The hypotheses are:

$$H_0: \quad \mu_e = \mu_s$$
$$H_a: \quad \mu_e > \mu_s$$

(c) We define ρ to be the correlation between levels of BMP and noggin in the brains of mice. The hypotheses are:

$$H_0: \quad \rho = 0$$
$$H_a: \quad \rho < 0$$

4.21 We define μ_m to be mean heart rate for males being admitted to an ICU and μ_f to be mean heart rate for females being admitted to an ICU. The hypotheses are:

$$H_0: \quad \mu_m = \mu_f$$
$$H_a: \quad \mu_m > \mu_f$$

4.23 We define ρ to be the correlation between systolic blood pressure and heart rate for patients admitted to an ICU. The hypotheses are:

$$H_0: \quad \rho = 0$$
$$H_a: \quad \rho > 0$$

Note: The hypotheses could also be written in terms of β, the slope of a regression line to predict one of these variables using the other.

4.25 We define μ to be the mean age of ICU patients. The hypotheses are:

$$H_0: \quad \mu = 50$$
$$H_a: \quad \mu > 50$$

4.27 (a) The population parameter of interest is the correlation, ρ, between number of children and household income from all households within the US. The hypotheses are:

$$H_0: \quad \rho = 0$$
$$H_a: \quad \rho \neq 0$$

(b) We are testing for a relationship, which means a correlation different from 0. So the larger correlation of 0.75 provides more evidence.

(c) Since we are simply looking for evidence of a relationship, the same amount of evidence is given from both 0.50 and -0.50. However they would give evidence of a relationship in opposite directions.

4.29 (a) The null hypothesis (H_0) is that Muriel has no ability to distinguish whether the milk or tea is poured first, and so her guesses are no better than random. The alternative hypothesis (H_a) is that Muriel's guesses for whether the milk or tea are poured first are better than random.

(b) Since there are only two possible answers (tea first or milk first), if she is guessing randomly, we expect her to be correct about half the time. The hypotheses are $H_0 : p = 0.5$ vs $H_a : p > 0.5$.

4.31 This analysis does not involve a test because there is no claim of interest. We would likely use a confidence interval to estimate the average.

4.33 This analysis does not include a test because from the information in a census, we can find exactly the true population proportion.

4.35 This analysis does not include a test because there is no claim of interest. The analysis would probably include a confidence interval to give an estimate of the average reaction time.

4.37 This analysis does not include a statistical test. Since we have all the information for the population, we can compute the proportion who voted exactly and see if it is greater than 50%.

4.39 (a) We define p_c to be the proportion supporting Candidate A after a phone call and p_f to be the proportion supporting Candidate A after a flyer. The hypotheses are:

$$H_0: \quad p_c = p_f$$
$$H_a: \quad p_c > p_f$$

(b) Answers will vary, but we need obviously more support when getting a call. Also remember that there are 100 voters in each group. One possible set of data is:

	Vote A	Not vote A
Phone Call	98	2
Flyer	50	50

(c) Answers will vary but, to have no evidence for H_a, we need to show *less* support when getting a call. One possible set of data is:

	Vote A	Not vote A
Phone Call	40	60
Flyer	60	40

(d) Answers will vary, but we need more support with a call, but only by a little. One possible set of data is:

	Vote A	Not vote A
Phone Call	52	48
Flyer	50	50

Section 4.2 Solutions

4.41 The sample mean, $\bar{x}$

4.43 The difference in the sample means, $\bar{x}_1 - \bar{x}_2$

4.45 The randomization distribution will be centered at 0.5, since $p = 0.5$ under H_0. Since $H_a : p < 0.5$, this is a left-tail test.

4.47 The randomization distribution will be centered at 0, since $\rho = 0$ under H_0. Since $H_a : \rho \neq 0$, this is a two-tailed test.

4.49 The randomization distribution will be centered at 0, since $p_1 - p_2 = 0$ under H_0. Since $H_a : p_1 > p_2$, this is a right-tail test.

4.51   (a) We see in the randomization distribution that there are no sample means even close to $\bar{x} = 250$ when the null hypothesis is true, so a sample mean this far out is (*iii*): extremely unlikely to ever occur using samples of this size.

  (b) We see in the randomization distribution that a sample mean of $\bar{x} = 305$ is not unusual when the null hypothesis is true, so a value this extreme is (*i*): reasonably likely to occur.

  (c) We see in the randomization distribution that a sample mean as extreme as $\bar{x} = 315$ is rare when the null hypothesis is true, but sample means as extreme occurred several times in this randomization distribution. A value this extreme is (*ii*): unusual but might occur occasionally.

4.53 This is a left-tail test, so in each case, we are estimating the proportion of the distribution to the left of the value.

  (a) $\hat{p} = 0.25$ is in the lower tail, but not very far, so the p-value is closer to 0.30 than 0.001.

  (b) $\hat{p} = 0.15$ is farther out in the lower tail so the p-value should be quite small, around 0.04.

  (c) $\hat{p} = 0.35$ is in the upper tail of the randomization distribution. Since this is a lower tail test ($H_a : p < 0.3$), the p-value for $\hat{p} = 0.35$ is the proportion of randomization samples with proportions below 0.35 which will be more than half the values. Thus the p-value is closer to 0.70 than 0.30.

4.55   (a) The figures showing the (two-tail) region beyond the observed statistic are shown below.

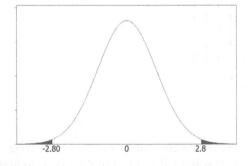 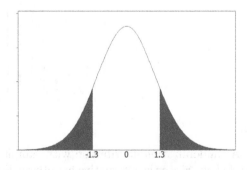

  (b) We see that $D = 2.8$ is farther out in the tail, so there is less area beyond it (and beyond -2.8). This means $D = 2.8$ has a smaller p-value and thus provides stronger evidence against H_0.

4.57 (a) We compute the difference in means: $D = \bar{x}_1 - \bar{x}_2 = 17.3 - 18.7 = -1.4$ and $D = \bar{x}_1 - \bar{x}_2 = 19.0 - 15.4 = 3.6$. The figures are shown below.

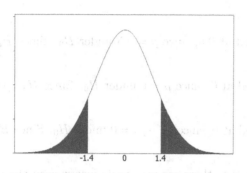 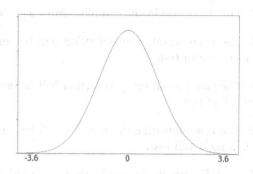

(b) The difference is larger in magnitude for $D = \bar{x}_1 - \bar{x}_2 = 19.0 - 15.4 = 3.6$ than $D = \bar{x}_1 - \bar{x}_2 = 17.3 - 18.7 = -1.4$. So $\bar{x}_1 = 19.0$ with $\bar{x}_2 = 15.4$ give a sample difference that is farther in a tail of the randomization distribution and provide stronger evidence against H_0.

4.59 A randomization distribution with sample proportions for 1000 samples of size 50 when $p = 0.50$ is shown below. Since the alternative hypothesis is $p > 0.5$, this is a right-tail test and we need to find how many of the randomization proportions are at (or above) the sample value of $\hat{p} = 0.60$. For this randomization distribution that includes 102 out of 1000 simulations so the p-value is 0.102. Answers will vary.

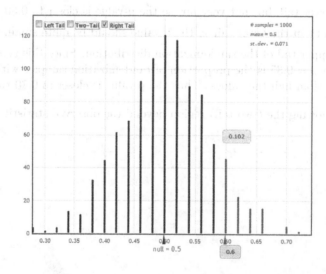

4.61 A randomization distribution with sample proportions for 1000 samples of size 200 when $p = 0.70$ is shown below. Since the alternative hypothesis is $p < 0.7$, this is a left-tail test and we need to find how many of the randomization proportions are at (or below) the sample value of $\hat{p} = 0.625$. For this randomization distribution that includes 9 out of 1000 simulations so the p-value is 0.009. Answers will vary.

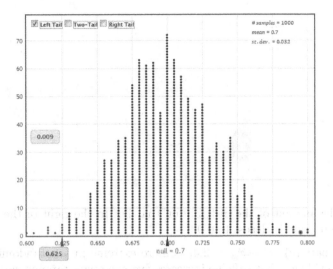

4.63 A randomization distribution with sample proportions for 1000 samples of size 100 when $p = 0.5$ is shown below. Since the alternative hypothesis is $p \neq 0.5$, this is a two-tail test. We need to find how many of the randomization proportions are at (or below) the sample value of $\hat{p} = 0.42$ and double to account for the other tail. For this randomization distribution there are 60 out of 1000 simulations beyond 0.42 so the p-value is $2 * 60/1000 = 0.120$. Answers will vary.

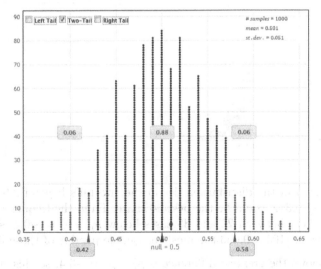

4.65 (a) Letting μ_c and μ_n represent the average tap rate of people who have had coffee with caffeine and without caffeine respectively, the null and alternative hypotheses are

$$H_0 : \mu_c = \mu_n$$
$$H_a : \mu_c > \mu_n$$

(b) This is a right-tail test so we shade the area to the right of the statistic 1.6. See the figure. The amount of the distribution that lies to the right of 1.6 is a relatively small portion of the entire graph. It is not as small as 0.03 and is not close to half the data so is not as large as 0.45 or 0.60. The p-value of 0.11 is the most reasonable estimate.

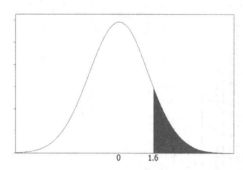 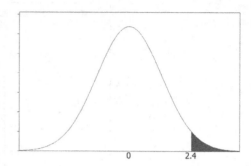

(c) See the figure. The amount of the distribution that lies to the right of the statistic $\bar{x}_c - \bar{x}_n = 2.4$ is very small, so the p-value is closest to 0.03.

(d) The difference in part (c), $\bar{x}_c - \bar{x}_n = 2.4$, is more extreme in the randomization distribution, so it provides stronger evidence that caffeine increases average finger tapping rate.

4.67 (a) Since this is a two-tail test, we shade *both* tails, first beyond ± 0.2, then beyond ± 0.4. See the figure.

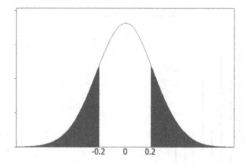 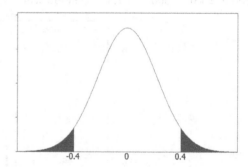

(b) Less than half the area of the distribution lies in the two tails beyond ± 0.2, but not a lot less than half, so the p-value of 0.392 would be the best estimate. From the frequencies in the histogram, there appear to be somewhere between 25 and 35 cases below $p_f - p_c = -0.4$, so doubling to account for two tails would make 66/1000=0.066 the most reasonable p-value of those listed.

(c) The statistic that shows the greatest difference is $p_f - p_c = -0.4$, so this statistic is likely to provide the strongest evidence that the methods are not equally effective.

4.69 (a) We are testing whether the proportion p of people who die from colon cancer after having polyps removed in a colonoscopy is less than 0.01. The hypotheses are:

$$H_0: \quad p = 0.01$$
$$H_a: \quad p < 0.01$$

(b) The sample proportion is $\hat{p} = 12/2602 = 0.0046$.

(c) We want to see how extreme the sample proportion of 0.0046 is on the randomization distribution. This is a left tail test, so we are interested in the number of dots to the left of 0.0046. It appears that there are four dots to the left of the sample statistic out of a total of 1000 dots, so

$$\text{p-value} = \frac{4}{1000} = 0.004.$$

4.71 (a) This is a test for a difference in means. If we use μ_C for mean hippocampus volume of the control people who have never played football and μ_F for mean hippocampus volume for football players, the hypotheses are:

$$H_0: \quad \mu_C = \mu_F$$
$$H_a: \quad \mu_C > \mu_F$$

(b) We see that $\bar{x}_C - \bar{x}_F = 7602.6 - 6459.2 = 1143.4$.

(c) This is a right-tail test. In the randomization distribution from 2000 simulations below, we see that there are no dots larger than the sample statistic of 1143.4, so the proportion more extreme is 0.000. Therefore, the p-value is approximately 0.000. Answers may vary, but p-value should be very small.

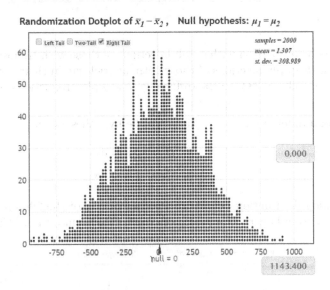

(d) This difference in means was more extreme than any that we saw in the simulated samples, so it is very unlikely that this difference is just the result of random chance.

4.73 (a) This is a test for a difference in proportions. If we use p_M for the proportion of US men who own a smartphone and p_W for the proportion of US women who own a smartphone, the hypotheses are:

$$H_0: \quad p_M = p_W$$
$$H_a: \quad p_M \neq p_W$$

(b) We see that $\hat{p}_M = 688/989 = 0.696$ and $\hat{p}_W = 671/1012 = 0.663$. The sample statistic is $\hat{p}_M - \hat{p}_W = 0.696 - 0.663 = 0.033$. We see that a larger proportion of men own smartphones.

(c) This is a two-tail test. The sample statistic 0.033 is in the right tail of the randomization distribution and we see that the proportion of the 3000 simulated samples more extreme than 0.033 in that direction is about 0.052. Since this is a two-tail test, we multiply by 2:

$$\text{p-value } = 2 \cdot 0.052 = 0.104.$$

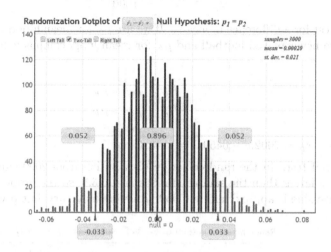

4.75 (a) If we let $\hat{p}_T$ be proportion with staph infections of those with triclosan in their system and $\hat{p}_N$ be the proportion with staph infections of those without triclosan in their systems, we have $\hat{p}_T = 24/37 = 0.649$ and $\hat{p}_N = 15/53 = 0.283$, so the difference in sample proportions is $\hat{p}_T - \hat{p}_N = 0.649 - 0.283 = 0.366$.

(b) This is a test for a difference in proportions, so the hypotheses are:

$$H_0: \quad p_T = p_N$$
$$H_a: \quad p_T > p_N$$

(c) In a randomization distribution below, we see that only 2 of the 2000 matched the oiginal difference of 0.366 and none were larger. The p-value from this distribution is $2/2000 = 0.001$.

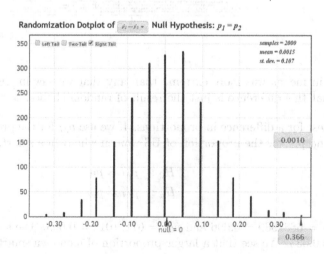

4.77 The randomization distribution represents samples chosen when H_0 is true. The area in a tail gives an estimate of the probability that a result as extreme (or more extreme) than the original sample should occur when H_0 is true, which is what the p-value measures.

4.79 (a) $X = 8$ is 5 above the expected count of 3. A point as far away in the other direction would be at -2. It's impossible to have negative values when counting how many students choose a number, so there could never be points that far away in the other direction.

(b) To find the p-value we double the proportion of randomization samples that are at or above $X = 8$. This only happened 3+1=4 times in 1000 randomizations, so the p-value $= 2 \cdot 4/1000 = 0.008$.

(c) The smallest possible lower tail p-value is 0.046 which would occur if none of the thirty students in the sample picked zero.

Section 4.3 Solutions

4.81 The smaller p-value, 0.04, provides stronger evidence against H_0.

4.83 The smaller p-value, 0.0008, provides stronger evidence against H_0.

4.85 Reject H_0, since p-value $= 0.0320 < 0.05$.

4.87 Do not reject H_0, since p-value $= 0.1145 \geq 0.05$.

4.89 The results are significant if the p-value is less than the significance level. A p-value of 0.2800 would mean the results are not significant at any of these levels.

4.91 The results are significant if the p-value is less than the significance level. A p-value of 0.0621 shows the results are significant at a 10% level, but not significant at 5% or 1% levels.

4.93 (a) II. 0.0571, less than 0.10, but not less than 0.05.

(b) I. 0.00008, smaller than any reasonable significance level.

(c) IV. 0.1753, larger than any reasonable significance level.

(d) III. 0.0368, less than 0.05, but not less than 0.01.

4.95 (a) The study found evidence to show a difference between the groups in spatial memory, which implies a low p-value for that test. The p-value for the test of spatial memory must be 0.0001. The study did not find a significant difference between the groups in amount of time exploring objects, which implies a relatively high p-value for that test. The p-value for the test of time exploring objects must be 0.7.

(b) The title implies causation, which is justified since the results come from a randomized experiment and we observed an effect.

4.97 (a) If the mean arsenic level is really 80 ppb, the chance of seeing a sample mean as high (or higher) than was observed in the sample from supplier A by random chance is only 0.0003. For supplier B, the corresponding probability (seeing a sample mean as high as B's when $\mu = 80$) is 0.35.

(b) The smaller p-value for Supplier A provides stronger evidence against the null hypothesis and in favor of the alternative that the mean arsenic level is higher than 80 ppb. Since it is very rare for the mean to be that large when $\mu = 80$, we have stronger evidence that there is too much arsenic in Supplier A's chickens.

(c) The chain should get chickens from Supplier B, since there is strong evidence that Supplier A's chicken have a mean arsenic level above 80 ppb which is unacceptable.

4.99 (a) This is a two-tailed test for a difference in means, so the hypotheses are:

$$H_0: \quad \mu_M = \mu_F$$
$$H_a: \quad \mu_M \neq \mu_F$$

where μ_M and μ_F represent mean nasal tip angle for males and females, respectively.

(b) Since the p-value is greater than 0.05, we do not reject H_0 and do not have evidence at a 5% level of a difference in average tip angle between males and females.

4.101   (a) The explanatory variable is whether or not antibiotics were given during the first year of life, and the response variable is whether or not the child was categorized as overweight at age 12. Both are categorical.

   (b) This is an observational study since the explanatory variable was not manipulated.

   (c) This is a one-tailed hypothesis test for a difference in proportions. Using p_A to represent the proportion of children who are overweight of those who have been given antibiotics in infancy and p_N to represent the proportion of children who are overweight of those who have not been given antibiotics, we have:

$$H_0: \quad p_A = p_N$$
$$H_a: \quad p_A > p_N$$

   (d) This is a test for a difference in proportions, so the sample statistic is $\hat{p}_A - \hat{p}_N$, where $\hat{p}_A$ represents the sample proportion of students overweight in the group that received antibiotics in infancy and $\hat{p}_N$ represents the sample proportion of students overweight in the group that did not receive antibiotics. We are told that $\hat{p}_A = 0.324$ and $\hat{p}_N = 0.182$, so the sample statistic is:

$$\hat{p}_A - \hat{p}_N = 0.324 - 0.182 = 0.142.$$

   (e) We are told that the p-value is 0.002 so the conclusion is the reject H_0. The p-value is very small so the evidence is strong.

   (f) No, we cannot conclude causation since the data come from an observational study.

4.103   (a) Students who think the drink is more expensive solve, on average, more puzzles than students who have a discounted price. The p-value is very small so the evidence for this conclusion is very strong.

   (b) If you price a product too low, customers might perceive it to be less effective or lower quality than it actually is.

4.105   (a) The hypotheses are $H_0 : p = 0.5$ vs $H_a : p \neq 0.5$ where p is the proportion of penalty shots that go to the right after a specific body movement.

   (b) The p-value (0.3184) is not small so we do not reject H_0. There is not evidence that this movement helps predict the direction of the shot, so there is no reason to learn to distinguish it.

   (c) The p-value (0.0006) is smaller than any common significance level so we reject H_0. The proportion of shots to the right after this movement is different from 0.50, so a goalie can gain an advantage by learning to distinguish this movement.

4.107   (a) Since they randomly applied treatments (filtered or polluted air) to mice, this is a randomized experiment.

   (b) We are testing whether or not a difference exists, in either direction, so this will be a two tailed test. This means a null hypothesis of no difference in mean insulin resistance between treatments $H_0 : \mu_{FA} = \mu_{PM}$, and an alternative that there is a difference $H_a : \mu_{FA} \neq \mu_{PM}$

   (c) Since the -4.4 is smaller than all the values in the 1,000 random simulations shown in the histogram of the exercise, the p-value is very small (≈ 0.000) and we will reject the null hypothesis.

   (d) The p-value is essentially zero.

   (e) Since the p-value is very small (essentially zero), we reject H_0 and conclude that there is strong evidence that mean insulin resistance scores are significantly affected by air pollution. There appears to be a strong connection between insulin resistance (diabetes) and air pollution.

4.109   (a) The small p-value (less than 0.01) gives strong evidence that Y-maze performance improves for mice after 7-10 days of exercise.

(b) The small p-value (less than 0.01) gives strong evidence that BMP levels decrease for mice after 2 or more days of exercise.

(c) The very small p-value (less than 0.001) gives very strong evidence that noggin levels increase for mice after 7-10 days of exercise.

(d) The strongest statistical effect is the test for noggin level which appears to have the smallest p-value.

(e) In mice that exercise, Y-maze performance improves after 7-10 days, BMP levels decrease after 2 or more days, and noggin levels increase after 7-10 days. Exercise appears to have a very significant positive effect on brain function in mice.

4.111   (a) Using p_E to represent the proportion of people who can solve the problem while getting electrical stimulation and p_N to represent the proportion of people that can solve the problem with no electrical stimulation, we are testing

$$H_0: \quad p_E = p_N$$
$$H_a: \quad p_E > p_N$$

Using a randomization distribution from StatKey or other technology, we use the sample statistic $\hat{p}_E - \hat{p}_N = 0.6 - 0.2 = 0.40$ to estimate the p-value as the area in the upper tail. For the distribution shown below, we see that the p-value is about 0.009. There is strong evidence to reject H_0 and conclude that people are more likely to solve the problem if they get electrical stimulation.

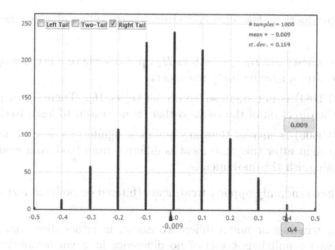

(b) Yes. The results are significant, and because this was a well-designed experiment, we can conclude that there is a causal relationship. Electrical stimulation of the brain appears to help people find fresh insight to solve a new problem.

4.113   (a) Let p be the true proportion that answer correctly. Under random guessing, the three choices are equally likely answers and therefore $p = 1/3$. So, the null and alternative hypotheses are $H_0: p = 1/3$ and $H_a: p \neq 1/3$.

(b) Since 5% of the 1005 people in the sample gave the correct answer, the count of correct answers is about 50. Using StatKey to generate a randomization distribution for proportions of samples of size 1005 when $p = 1/3$ shows no values anywhere near the $\hat{p} = 0.05$ in the original sample. Thus the p-value ≈ 0.000.

(c) The p-value is less than any commonly used significance threshold. Therefore, we reject the null hypothesis and conclude that U.S. citizens give the correct answer less often than would be expected if they were randomly guessing.

4.115 (a) The relevant parameters are the average difference between actual and scheduled arrival time for *Delta* (μ_D) and *United* (μ_U). The null hypothesis is $\mu_D = \mu_U$ and the alternative hypothesis is $\mu_D \neq \mu_U$.

(b) The sample mean for the *Delta* flights is -2.6 minutes (so 2.6 minutes early), and the sample mean for the *United* flights is 9.8 minutes late. The difference in means is $\overline{x}_D - \overline{x}_U = -2.6 - 9.8 = -12.4$.

(c) Generating 10,000 samples in a randomization distribution with *StateKey* we don't find a single sample with a difference as extreme as $\overline{x}_D - \overline{x}_U = -12.4$ so p-value ≈ 0.

(d) Since the p-value is very small we reject the null hypothesis in favor of the alternative. We have strong evidence the mean difference between actual and schedule arrival times is smaller for Delta than United.

4.117 (a) $H_0 : \mu_i = \mu_u$, $H_a : \mu_i > \mu_u$, where μ_i and μ_u are the mean number of twitches for lizards from invaded and uninvaded habitats (respectively).

(b) In the randomization distribution below, only 1 out of 5000 simulated statistics is as extreme as the observed statistic of $\overline{x}_1 - \overline{x}_u = 2.75 - 1.1 = 1.65$, so the p-value is 0.0002.

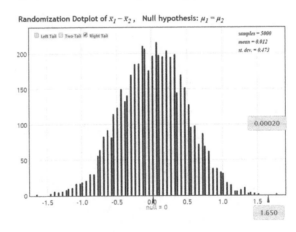

(c) This is a small p-value so we reject H_0. We have strong evidence that the mean number of twitches when lizards from an invaded habitat encounter fire ants is larger than for lizards from an uninvaded habitat.

4.119 (a) No, case-control studies can never be used to infer causality because they are observational studies; neither variable is manipulated by the researchers.

(b) Choosing controls similar to cases is a way of eliminating some confounding variables. For example, if each control was the same age, sex, and socioeconomic status as a case, we can rule these out as confounding variables. However, because we cannot ensure that controls are similar to cases regarding *all* potential confounding variables, we still cannot make conclusions about causality.

(c) $\hat{p}_1 - \hat{p}_2 = 136/262 - 220/522 = 0.098$

(d) We use technology to generate the randomization distribution below, and find the proportion of statistics in the upper tail with a difference of at least 0.098 or higher. This yields a p-value of 0.0042.

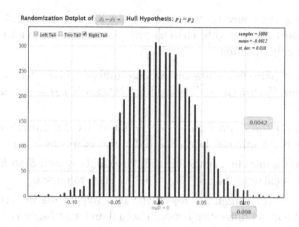

(e) This p-value is very small, and so results this extreme would be very unlikely to happen just by random chance. This provides convincing evidence that there is an association between owning a cat as a child and developing schizophrenia. Note: This does not imply there is a causal relationship.

Section 4.4 Solutions

4.121   (a) In a randomization distribution below (on the left), we see that the p-value is 0.167. The results are not significant when $n = 50$.

(b) In a randomization distribution below (on the right), we see that the p-value is 0.0002. The results are significant and the evidence is very strong when $n = 500$.

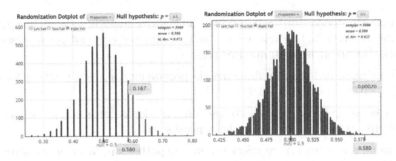

We find the strongest evidence for the alternative hypothesis with the largest sample size, when $n = 500$.

4.123   (a) In a randomization distribution below (on the left), we see that the p-value is 0.272. The results are not significant when $n_1 = n_2 = 20$.

(b) In a randomization distribution below (in the middle), we see that the p-value is 0.0012. The results are significant when $n_1 = n_2 = 200$.

(c) In a randomization distribution below (on the right), we see that the p-value is 0.000. With these sample sizes, there are no simulated samples anywhere near as extreme as 0.15. The results are significant and the evidence is very strong when $n_1 = n_2 = 2000$.

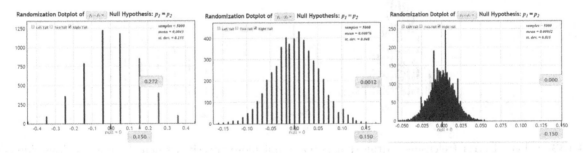

We find the strongest evidence for the alternative hypothesis with the largest sample size, when $n_1 = n_2 = 2000$.

4.125 If the null hypothesis is true, we are still likely to find significance in 1% of the tests, which is $0.01 \times 300 = 3$ of the tests.

4.127 If the null hypothesis is true, we are still likely to find significance in 5% of the tests, which is $0.05 \times 800 = 40$ of the tests.

4.129   (a) We know the two-tailed p-value is less than 0.05, but can't tell if it's also less than 0.01, so we can't make a decision for the 1% test.

(b) If we reject H_0 at a 5% level, the p-value is less than 0.05, so it is also less than 0.10 and thus we would also reject H_0 at a 10% level.

(c) If the sample correlation $r > 0$, its p-value in an upper tail test would be one half of the two-tailed p-value. Since the two-tailed p-value is less than 0.05, half of it will be even smaller and thus also less than 0.05. Since the p-value for the one-tailed test is less than 5%, H_0 will be rejected at a 5% level, so the conclusion is valid. (Note, however, if the sample $r < 0$ (i.e. is in the lower tail) then it's upper tail p-value would be more than 0.50 and show no evidence to reject $H_0 : \rho = 0$ in favor of $H_a : \rho > 0$.)

4.131 (a) The null hypothesis is $p_D = p_U$ and the alternative hypothesis is $p_D \neq p_U$, where p_D represents the proportion of *Delta* flights that are at least 30 minutes late and p_U represents the proportion of *United* flights that are at least 30 minutes late.

(b) The statistic being recorded is the difference in proportions, $\hat{p}_D - \hat{p}_U$, which for our observed sample is $\frac{67}{1000} - \frac{160}{1000} = 0.067 - 0.160 = -0.093$.

(c) Generating 5000 samples in *StatKey* (below left) we don't get a single sample with a difference as extreme as 0.067 - 0.160 = -0.093 so we have p-value ≈ 0

(d) Since the p-value is less than α we reject the null hypothesis in favor of the alternative, meaning that we do find evidence that there is a significant difference in the proportion of flights arriving more than 30 minutes late and that proportion this late is smaller for Delta than United.

(e) With the new sample sizes , we would not come to the same conclusion. We see in a randomization distribution (below right) that the p-value is $2 * 0.058 = 0.116$, so we would fail to reject the null hypothesis. With this sample, we don't have enough evidence to claim that the proportion of delayed flights is different.

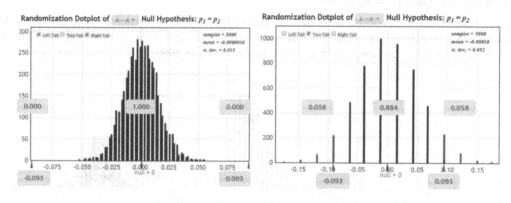

4.133 A Type I error (releasing a drug that is really not more effective) when there are serious side effects should be avoided, so it makes sense to use a small significance level such as $\alpha = 0.01$.

4.135 A Type I error (saying your average Wii bowling score is higher than a friend's, when it isn't) is not very serious, so a large significance level such as $\alpha = 0.10$ will make it easier to see any difference.

4.137 A Type I error (getting people to take the supplements when they don't help) is not serious if there are no harmful side effects, so a large significance level, such as $\alpha = 0.10$, will make it easier to see any benefit of the supplements.

4.139 Type I error: Release a drug that is really not more effective. Type II error: Fail to show the drug is more effective , when actually it is. Personal opinions will vary on which is worse.

4.141 Type I error: Find your average Wii bowling score is higher than a friend's, when actually it isn't. Type II error: The sample does not contain enough evidence to conclude that your average Wii bowling score is better than your friend's mean, when actually it is better. Personal opinions will vary on which is worse.

4.143 Type I error: Conclude people should take the supplements when they actually don't help. Type II error: Fail to detect that the supplements are beneficial, when they actually are. Personal opinions will vary on which is worse.

4.145 (a) If the sample shows significant results we reject H_0. If that conclusion is right, we have not made an error. If that conclusion is wrong (i.e., H_0 is true) we've made a Type I error.

  (b) If the sample shows insignificant results we do not reject H_0. If that conclusion is right, we have not made an error. If that conclusion is wrong (i.e., H_0 is false) we've made a Type II error.

  (c) We would need to know the actual value of the parameter for the population to verify if we made the correct decision or an error, and if we knew the actual value of the parameter we would not need to do any statistical inference.

4.147 (a) We let μ_I and μ_C represent the mean score on the HRSIW subtest for kindergartners who get iPads and kindergartners who don't get iPads, respectively. The hypotheses are:

$$H_0: \quad \mu_I = \mu_C$$
$$H_a: \quad \mu_I > \mu_C$$

  (b) Since the p-value (0.006) is very small, we reject H_0. There is evidence that the mean score on the HRSIW subtest for kindergartners with iPads is higher than the mean score for kindergartners without iPads. The results are statistically significant.

  (c) The school board member could be arguing whether a 2 point increase in the mean score on one subtest really matters very much (is *practically* significant). Even though this is a statistically significant difference, it might not be important enough to justify the considerable cost of supplying iPad's to all kindergartners.

4.149 (a) If there is no effect due to food choices and $\alpha = 0.01$, a Type I error should occur on about 1% of tests. We see that 1% of 133 is 1.33, so we would expect one or two tests to show significant evidence just by random chance.

  (b) When testing more than 100 different foods, making at least one Type I error is fairly likely.

  (c) No, even if the proportion of boys is really higher for mothers who eat breakfast cereal, the data were obtained from an observational study and not an experiment. A headline that implies eating breakfast cereal will *cause* an increase in the chance of conceiving a boy is not appropriate without doing an experiment where cereal habits are controlled.

4.151 (a) We should definitely be less confident. If the authors conducted 42 tests, it is likely that some of them will show significance just by random chance even if massage does not have any significant effects. It is possible that the result reported earlier just happened to one of the random significant ones.

  (b) Since none of the tests were significant, it seems unlikely that massage affects muscle metabolites.

  (c) Now that we know that only eight tests were testing inflammation, and that four of those gave significant results, we can be quite confident that massage does reduce muscle inflammation after exercise. It would be very surprising to see four p-values (out of eight) less than 5% if there really were no effects at all.

4.153 (a) We create a randomization distribution for the difference in proportions, and calculate the p-value as the proportion of simulated statistics at least as high as the sample statistics, $\hat{p}_1 - \hat{p}_2 = 0.0091$, yielding a p-value of 0.106.

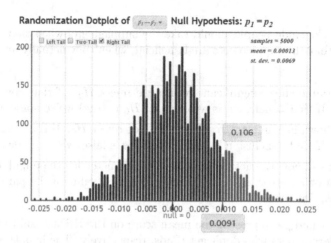

(b) The p-value of 0.106 is not smaller than $\alpha = 0.05$, so we do not reject the null. We do not have sufficient evidence to conclude that mate choice improves offspring fitness in fruit flies.

(c) We create a randomization distribution, and calculate the p-value as the proportion of simulated statistics at least as high as the sample mean difference, $\bar{x} = 1.82$ flies, yielding a p-value of 0.204.

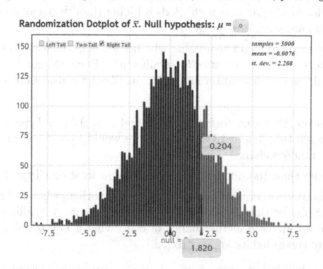

(d) The p-value of 0.204 is not smaller than $\alpha = 0.05$, so we do not reject the null. We do not have sufficient evidence to conclude that mate choice improves offspring fitness in fruit flies.

(e) The follow-up study did not find significant results, so if mate choice really does improve offspring fitness in fruit flies, a Type II error would have been made.

(f) The original study did find significant results, so if mate choice really does not improve offspring fitness in fruit flies, a Type I error would have been made.

Section 4.5 Solutions

4.155 (a) We use a confidence interval since we are estimating the proportion of voters who support the proposal and there is no claimed parameter value to test.

(b) We use a hypothesis test since we are testing the claim $H_0 : p_h = p_a$ vs $H_a : p_h > p_a$.

(c) We use a hypothesis test to test the claim $H_0 : p = 0.5$ vs $H_a : p > 0.5$.

(d) Inference is not relevant in this case, since we have information on the entire populations.

4.157 (a) Since the null $\mu = 15$ is in the 95% confidence interval, we do not reject H_0 using a 5% significance level.

(b) Since the null $\mu = 15$ is outside of the 95% confidence interval, we reject H_0 using a 5% significance level.

(c) Since the null $\mu = 15$ is in the 90% confidence interval, we do not reject H_0 using a 10% significance level.

4.159 (a) Since the null $\rho = 0$ is outside of the 95% confidence interval, we reject H_0 using a 5% significance level. The interval contains only positive values for the correlation, providing evidence that the correlation for the population is positive.

(b) Since the null $\rho = 0$ is outside of the 90% confidence interval, we reject H_0 using a 10% significance level. The interval contains only negative values for the correlation, providing evidence that the correlation for the population is negative.

(c) Since the null $\rho = 0$ is in the 99% confidence interval, we do not reject H_0 using a 1% significance level.

4.161 (a) Since the null $p = 0.5$ is inside the 95% confidence interval, we do not reject H_0 using a 5% significance level.

(b) Since the null $p = 0.75$ is above the 95% confidence interval, we reject H_0 using a 5% significance level.

(c) Since the null $p = 0.4$ is below the 95% confidence interval, we reject H_0 using a 5% significance level.

4.163 (a) Since the null $p_1 - p_2 = 0$ is below the 90% confidence interval, we reject H_0 using a 10% significance level.

(b) Since the null $p_1 - p_2 = 0$ is below the 90% confidence interval, we reject $H_0 : p_1 = p_2$ vs the two-tailed alternative $H_a : p_1 \neq p_2$ using a 10% significance level. That means that the p-value for the two-tailed test is less than $\alpha = 0.10$. Since the confidence interval contains only positive values, we also know that $\hat{p}_1 - \hat{p}_2 > 0$. Thus the original difference is in the upper tail of a randomization distribution for the test and the proportion beyond it (half the p-value) must be less than 0.05. This means the upper-tail test would lead to rejecting H_0 at a 5% significance level.

(c) The confidence interval having positive endpoints shows that $\hat{p}_1 > \hat{p}_2$ in the original sample, so there is no reasonable evidence in the direction of the alternative $H_a : p_1 < p_2$. We do not reject H_0.

4.165 (a) No treatments were controlled so this is an observational study.

(b) The proportion of melanomas on the left side is $\hat{p} = 31/42 = 0.738$.

(c) We are 95% sure that between 57.9% and 86.1% of all melanomas occur on the left side.

(d) The hypotheses are $H_0 : p = 0.5$ vs $H_a : p > 0.5$, where p is the proportion of melanomas on the left.

(e) This is a one-tailed test since the question asks about the proportion on the left being more than on the right.

(f) Reject H_0, since the plausible values for the proportion are 0.579 to 0.862 which are all above the hypothesized 0.50.

(g) Since the p-value is small (less than any reasonable significance level), we reject H_0. This provides strong evidence that melanomas are more likely to occur on the left side than the right side.

(h) No, since the data were not collected with an experiment we cannot infer a cause and effect relationship between these variables.

4.167 (a) Since 0.5 is not in the 95% confidence interval, it is not a plausible value for the population proportion, so we reject H_0. There is evidence, at the 5% level, that the proportion with both partners reporting being in a relationship on Facebook is more than 0.5.

(b) Since 0.5 is in the 95% confidence interval, it is a plausible value for population proportion, so we do not reject H_0. There is not enough evidence, at the 5% level, to conclude that the proportion with both showing a partner in their Facebook profile pictures is different from 0.5.

4.169 (a) The proportion of home wins in the sample is $\hat{p} = 70/120 = 0.583$. Using *StatKey* or other technology, we construct a bootstrap distribution of sample proportions such as the one below. We see that a 90% confidence interval in this case goes from 0.508 to 0.658. We are 90% confident that the home team will win between 50.8% and 65.8% of soccer games in this league.

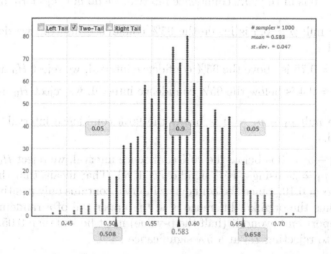

(b) To test if the proportion of home wins differs from 0.5, we use $H_0 : p = 0.5$ vs $H_a : p \neq 0.5$.

(c) Since 0.5 is not in the interval in part (a), we reject H_0 at the 10% level. The proportion of home team wins is not 0.5, at a 10% level.

(d) We create a randomization distribution (shown below) of sample proportions when $n = 120$ using $p = 0.5$. Since this is a two-tailed test, we have p-value = 2(0.041) = 0.082.

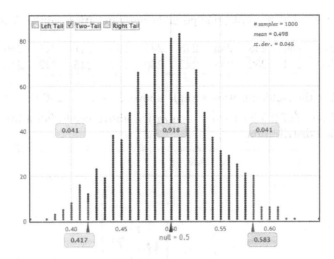

(e) At a 10% significance level, we reject H_0 and find that the proportion of home team wins is different from 0.5. Yes, this does match what we found in part (c).

(f) The confidence interval shows an estimate for the proportion of times the home team wins, which the p-value does not give us. The p-value gives a sense for how strong the evidence is that the proportion of home wins differs from 0.5 (only significant at a 10% level in this case).

(g) The bootstrap and randomizations distributions are similar, except that the bootstrap proportions are centered at the original $\hat{p} = 0.583$, while the randomization proportions are centered at the null hypothesis, $p = 0.5$.

4.171 (a) We are testing $H_0 : \mu = 10$ vs $H_a : \mu \neq 10$, where μ represents the mean longevity, in years, of all mammal species. The mean longevity in the sample is $\overline{x} = 13.15$ years. We use *StatKey* or other technology to create a randomization distribution such as the one shown below. For this two-tailed test, the proportion of samples beyond $\overline{x} = 13.15$ in this distribution gives a p-value of $2 \cdot 0.005 = 0.010$. We have strong evidence that mean longevity of mammal species is different from 10 years.

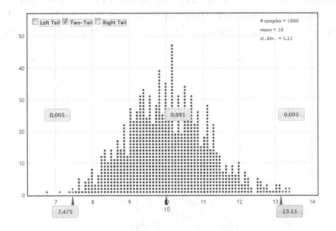

(b) Since, for a 5% significance level, we reject $\mu = 10$ as a plausible value of the population mean in the hypothesis test in part (a), 10 would not be included in a 95% confidence interval for the mean longevity of mammals.

4.173   (a) Answers vary. For example, one possible randomization sample is shown below

caffeine	244	250	248	246	248	245	246	247	248	246	$\bar{x}_c = 246.8$
no caffeine	250	244	252	248	242	250	242	245	242	248	$\bar{x}_n = 246.3$

(b) Answers vary. For the randomization sample above, $\bar{x}_c - \bar{x}_n = 246.8 - 246.3 = 0.5$.

(c) The sample difference of 0.5 for the randomization above would fall a bit to the right of the center of the randomization distribution.

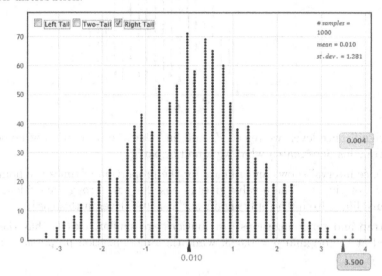

4.175 We use technology to create randomization samples by repeatedly sampling six values with replacement (after subtracting 11 from each value in the original sample to match the null hypothesis that $\mu = 80$). We collect the means for 1000 such samples to form a randomization distribution such as the one shown below. To find the p-value for this right-tailed alternative we count how many of the randomization means exceed our original sample mean, $\bar{x} = 91.0$. For the distribution below this is 115 of the 1000 samples, so the p-value = 0.115. This is not enough evidence (using a 5% significance level) to reject $H_0 : \mu = 80$. The chain should continue to order chickens from this supplier (but also keep testing the arsenic level of its chickens).

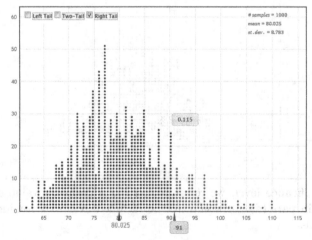

4.177 (a) Define ρ to be the correlation between number of hurricanes and year. We have $H_0 : \rho = 0$ (no association between hurricanes and years) and $H_a : \rho > 0$ (number of hurricanes tends to increase as years increase).

(b) We could create note cards with the number of hurricanes each year, shuffle them, then randomly assign the number of hurricane note cards to each of the years from 1914 to 2014. Calculate the correlation between number of hurricanes and years for the simulated sample and then repeat the process thousands of times.

4.179 We are testing whether the correlation ρ is positive, where ρ represents the correlation between the malevolence rating of uniforms in the National Hockey League and team z-scores for the number of penalty minutes. The hypotheses for the test are

$$H_0 : \quad \rho = 0$$
$$H_a : \quad \rho > 0$$

Using the data for *NHL_Malevolence* and *ZPenMin* in the dataset **MalevolentUniformsNHL** and using either StatKey or other technology, we match the null assumption that the correlation is zero by randomly assigning the values for one of the variables to the values of the other variable and compute the correlation of the resulting simulated data. We do this 1000 times and collect these simulated correlation statistics into a randomization dotplot (shown below). To see how extreme the original sample correlation of $r = 0.521$ is we find the proportion of simulated samples that have a correlations of 0.521 or larger. For the distribution below, this includes 18 of the 1000 randomizations in the right tail, giving a p-value $18/1000 = 0.018$. Answers will vary for other randomizations. Using a 5% significance level, we reject H_0 and conclude that the malevolence of hockey uniforms is positively correlated with the number of minutes a team is penalized. The results are significant at a 5% level but not at a 1% level.

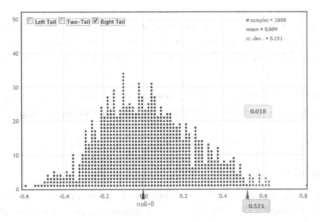

4.181 (a) Using *StatKey* or other technology we create randomization samples by randomly scrambling the desipramine/placebo group assignments and recording the difference (desipramine relapses - placebo relapses) for each randomization sample. We could record the differences in the counts for many randomizations (as in the randomization distribution below) or use the differences in the proportion who relapse in each group. The difference in counts for the original sample is $D = 10 - 20 = -10$. Since the alternative is left-tailed (looking for evidence of fewer relapses with desipramine) we count the number of randomization samples that give a difference of -10 or less. That is just 4 cases in the distribution below, giving a p-value of 0.004. This small p-value shows strong evidence to reject H_0 and conclude that desipramine works better than a placebo (gives a smaller proportion of patients relapsing) when treating cocaine addiction.

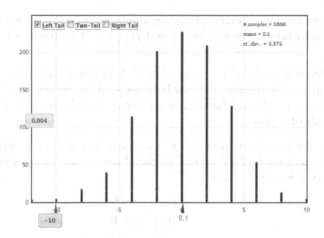

(b) The p-value for testing desipramine vs a placebo (0.004) is much smaller than the p-value for testing lithium vs a placebo (0.345, which is not significant). There is stronger evidence that desipramine is effective for treating cocaine addiction than there is for lithium.

4.183 (a) We are interested in whether pulse rates are higher on average than lecture pulse rates, so our hypotheses are

$$H_0: \quad \mu_Q = \mu_L$$
$$H_a: \quad \mu_Q > \mu_L$$

where μ_Q represents the mean pulse rate of students taking a quiz in a statistics class and μ_L represents the mean pulse rate of students sitting in a lecture in a statistics class. We could also word hypotheses in terms of the mean difference $D = Lecture\ pulse - Quiz\ pulse$ in which case the hypotheses would be $H_0: \mu_D = 0$ vs $H_a: \mu_D > 0$.

(b) We are interested in the difference between the two pulse rates, so an appropriate statistic is $\overline{x}_D$, the differences ($D = Lecture - Quiz$) for the sample. For the original sample the differences are:

$$+2, \quad -1, \quad +5, \quad -8, \quad +1, \quad +20, \quad +15, \quad -4, \quad +9, \quad -12$$

and the mean of the differences is $\overline{x}_D = 2.7$.

(c) Since the data were collected as pairs, our method of randomization needs to keep the data in pairs, so the first person is labeled with $(75, 73)$, with a difference in pulse rate of 2. As long as we keep the data in pairs, there are many ways to conduct the randomization. In every case, of course, we need to make sure the null hypothesis (no difference) is met and we need to keep the data in pairs.

One way to do this is to sample from the pairs with replacement, but randomly determine the order of the pair (perhaps by flipping a coin), so that the first pair might be $(75, 73)$ with a difference of 2 or might be $(73, 75)$ with a difference of -2. Notice that we match the null hypothesis that the quiz/lecture situation has no effect by assuming that it doesn't matter – the two values could have come from either situation. Proceeding this way, we collect 10 differences with randomly assigned signs (positive/negative) and compute the average of these differences. That gives us one simulated statistic.

A second possible method is to focus exclusively on the differences as a single sample. Since the randomization distribution needs to assume the null hypothesis, that the mean difference is 0, we can subtract 2.7 (the mean of the original sample of differences) from each of the 10 differences, giving the values

$$-0.7, \quad -3.7, \quad 2.3, \quad -10.7, \quad -1.7, \quad 17.3, \quad 12.3, \quad -6.7, \quad 6.3, \quad -14.7.$$

Notice that these values have a mean of zero, as required by the null hypothesis. We then select samples of size ten (with replacement) from the adjusted set of differences (perhaps by putting the 10 values on cards or using technology) and compute the average difference for each sample.

There are other possible methods, but be sure to use the paired data values and be sure to force the null hypothesis to be true in the method you create!

(d) Here is one sample if we randomly assign $+/-$ signs to each difference:

$$+2, \quad +1, \quad -5, \quad -8, \quad +1, \quad -20, \quad -15, \quad +4, \quad +9, \quad +12 \qquad \Rightarrow \overline{x}_D = -1.9$$

Here is one sample drawn with replacement after shifting the differences.

$$-6.7, \quad -1.7, \quad 6.3, \quad 2.3, \quad -3.7, \quad -1.7, \quad -0.7, \quad 17.3, \quad 2.3, \quad -0.7 \qquad \Rightarrow \overline{x}_D = 1.3$$

(e) Neither of the statistics for the randomization samples in (d) exceed the value of $\overline{x}_D = 2.7$ from the original sample, but your answers will vary for other randomizations.

4.185 (a) Sampling 100 responses (with replacement) from the original sample is inappropriate, since the samples are taken from a "population" where the proportion is 0.76, so it doesn't match $H_0 : p = 0.8$.

(b) Sampling 100 responses (with replacement) from a set consisting of 8 correct responses and 2 incorrect responses is appropriate, since $p = 0.8$ in this population and we are taking the same size samples as the original data.

4.187 We are testing $H_0 : \mu_F = \mu_M$ vs $H_0 : \mu_F \neq \mu_M$, where μ represents the average hours of exercise in a week. The difference for the original sample is $\overline{x}_F - \overline{x}_M = 9.40 - 12.40 = -3.00$. We need to see how many randomizations samples give mean differences this small (or even more extreme), then double the count to account for the other tail of the randomization distribution since this is a two-tailed alternative. Note, since this is a two-tailed test, we can just as easily use $\overline{x}_M - \overline{x}_F = 12.40 - 9.40 = +3.00$ as the difference for the original sample and look for more extreme values in the other tail.

(a) We randomly scramble the gender labels (M and F) and pair them with the actual exercise times, then find the mean exercise time within each randomly assigned group. (*StatKey* tip: Use "Reallocate" as the method of randomization.) For each randomization we record the difference in means $\overline{x}_F - \overline{x}_M$. Repeating this for 1000 randomizations produces a distribution such as the one shown below. To find the p-value we count the number of differences at (or below) the original sample difference of -3.00 and double the count to account for the other tail. In this case we have p-value= $2 * 99/1000 = 0.198$.

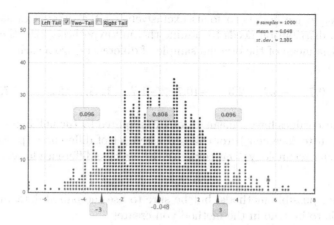

(b) Depending on technology, for this approach it may help to create two separate samples, one for the females and one for the males. Add 1.2 to each of the exercise values in the female sample and subtract 1.8 from each of the exercise values in the male sample to create new variables that have the same mean (10.6) in each group. Take separate samples (with replacement) — size 30 from the females and size 20 from the males — and find the mean exercise value in each sample. To find the randomization distribution we collect 1000 such differences. Depending on technology it might be easier to generate 1000 means for each gender and then subtract to get the 1000 differences. (*StatKey* tip: Use "Shift Groups" as the method of randomization.) One set of 1000 randomization differences in means is shown below. This distribution has 98 sample differences less than (or equal to) the original difference of -3.00, so the two-tailed p-value= $2 \cdot 98/1000 = 0.196$.

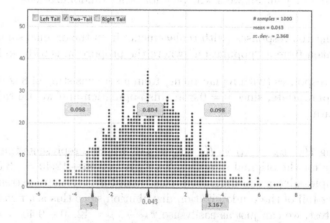

(c) We sample 30 values (with replacement) from the original sample of all 50 exercise amounts to simulate a new sample females and do the same for 20 males. Since both samples are drawn from the same set, we satisfy $H_0 : \mu_F = \mu_M$. (*StatKey* tip: Use "Combine Groups" as the method of randomization.) Compute the difference in means, $\overline{x}_F - \overline{x}_M$, for the new samples. One set of 1000 randomizations using this method is shown below. We see that 99 differences in means are at (or below) the original -3.00, so the two-tailed p-value= $2 \cdot 99/1000 = 0.198$.

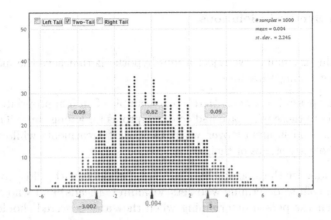

Note that that the randomization distributions and p-values produced by each method are similar. In each case the p-value is not small and we have insufficient evidence to conclude that there is difference in mean exercise time between female and male students.

Unit B: Essential Synthesis Solutions

B.1   (a) The p-value is small so we reject a null hypothesis that says the mean recovery time is the same with or without taking Vitamin C.

(b) There are many possible answers. Here's one example of an inappropriate data collection method: Give Vitamin C only to those who have had cold symptoms for a long time. They may have shorter recovery times since the cold is almost over when they start treatment, while those not getting Vitamin C might be at the early stages of their colds.

(c) We need to randomize the assignment of subjects (students with colds) to the two groups. For example, we could flip a coin to determine who gets Vitamin C (heads) and who gets a placebo (tails). Neither the subjects nor the person determining when they are recovered should know which group they are in.

(d) The small p-value indicates we should reject $H_0 : \mu_c = \mu_{nc}$ in favor of the alternative $H_0 : \mu_c < \mu_{nc}$, where μ denotes the mean recovery time from a cold. Thus we have strong evidence that large doses of Vitamin C help reduce the mean time students need to recover from a cold.

B.3   (a) The hypotheses are $H_0 : \mu_{dc} = \mu_w$ vs $H_a : \mu_{dc} > \mu_w$, where μ_{dc} and μ_w are the mean calcium loss after drinking diet cola and water, respectively.The difference in means for the sample is $\bar{x}_{dc} - \bar{x}_w = 56.0 - 49.125 = 6.875$. We use *StatKey* or other technology to construct a randomization distribution, such as the one shown below, for this difference in means test. We find the p-value in this upper-tail test by finding the proportion of the distribution above the sample difference in means of 6.875. For the randomization distribution below this gives a p-value of 0.005 and strong evidence to reject the null hypothesis. We conclude that mean calcium loss for women is higher when drinking diet cola than when drinking water.

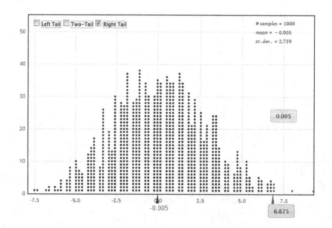

(b) Since we found a significant difference in part (a), we find a confidence interval for the difference in means $\mu_{dc} - \mu_w$ using a bootstrap distribution and either percentiles or the $\pm 2 \cdot SE$ method. For the bootstrap distribution shown below, we see that a 95% confidence interval using percentiles goes from 2.88 to 10.75. We are 95% sure that women who drink 24 ounces of diet cola will increase calcium excretion, on average, between 2.875 and 10.75 milligrams when compared to drinking water. Using $\pm 2 \cdot SE$, the 95% confidence interval is $6.875 \pm 2 \cdot 2.014 = 6.875 \pm 4.028 = (2.85, 10.90)$.

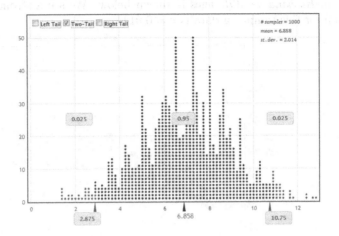

B.5 (a) Roommates are assigned at random, so whether a student has a roommate with a videogame or not is determined at random.

(b) The hypotheses are $H_0 : \mu_v = \mu_n$ vs $H_a : \mu_v < \mu_n$, where μ_v and μ_n are the mean GPA for students whose roommates do and do not bring a videogame, respectively.

(c) At a 5% level, we reject H_0. The mean GPA is lower when a roommate brings a videogame.

(d) Negative differences indicate $\mu_v - \mu_n < 0$ which means $\mu_v < \mu_n$, a lower mean GPA when a roommate brings a videogame. We are 90% sure that students with a roommate who brings a videogame have a mean GPA between 0.315 and 0.015 less than the mean GPA of students whose roommates don't bring a videogame.

(e) At a 5% significance level, we do not reject H_0 when the p-value=0.068. There is not (quite) enough evidence to show that mean GPA among students who don't bring a video game is lower when their roommate does bring one.

(f) At a 5% significance level we reject H_0 when the p-value=0.026. There is enough evidence to show that mean GPA among students who bring a video game is lower when their roommate also brings one.

(g) The effect (reducing mean GPA when a roommate brings a videogame) is larger for students who bring a videogame themselves. Perhaps this makes sense since students who bring a videogame are already predisposed to get distracted by them.

(h) For students who bring a video game to college, we are 90% sure that their mean GPA is lower by somewhere between 0.526 and 0.044 points if their roommate also brings a videogame than if their roommate does not bring a videogame. This interval is similar to the one found in part (d) but is farther in the negative direction.

(i) Having more videogames in the room tends to be associated with lower mean GPA.

(j) There are many possible answers. One possible additional test is to ignore the roommate completely and see if mean GPA in the first semester is lower for students who bring a videogame to college than for students who do not bring one.

B.7   (a) We expect married couples to tend to have similar ages, so we expect a positive correlation between husband and wife ages.

(b) A scatterplot of *Husband* vs *Wife* ages is shown below. We see a strong positive, linear association. The correlation for this sample of data is $r = 0.914$.

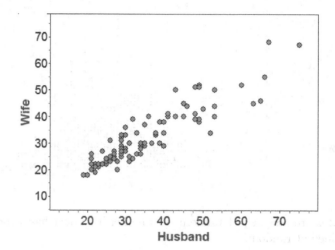

(c) To find a confidence interval for this correlation, we sample (with replacement) from the original data and compute the correlation between husband and wife ages for each bootstrap sample of 105 couples. We repeat this process to generate 5000 values in a bootstrap distribution such as the one shown below. From the percentiles of this distribution of bootstrap correlations, we find a 95% confidence interval to be from 0.877 to 0.945. We are 95% sure that the correlation between husband and wife ages for all recent marriages in this jurisdiction is between 0.877 and 0.945.

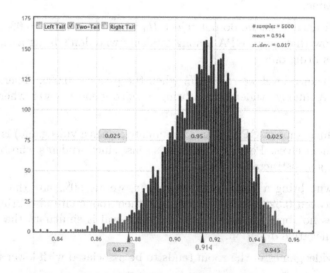

(d) Although we have evidence of a strong, positive correlation between the ages of husbands and wives, the correlation contains no information to help with the previous exercise of deciding whether husbands or wives tends to be older.

Unit B: Review Exercise Solutions

B.9 We are estimating p, the proportion of all US adults who own a laptop computer. The quantity that gives the best estimate is $\hat{p}$, the proportion of our sample who own a laptop computer. The best estimate is $\hat{p} = 1238/2252 = 0.55$. Since the true proportion is unknown, our best estimate for the proportion comes from our sample. We estimate that 55% of all US adults own a laptop computer.

B.11 (a) The relevant population is all American adults, and the parameter we are estimating is p, the proportion of all American adults who believe that violent movies lead to more violence in society. The best point estimate is $\hat{p} = 0.57$.

(b) A 95% confidence interval is

$$
\begin{array}{ccc}
\text{Point Estimate} & \pm & \text{Margin of Error} \\
0.57 & \pm & 0.03 \\
0.54 & \text{to} & 0.60
\end{array}
$$

We are 95% confident that the proportion of all American adults to believe that violent movies lead to more violence in society is between 0.54 and 0.60.

B.13 (a) This is a population proportion so the correct notation is p. Using the data in **Hollywood-Movies**, we have $p = 177/691 = 0.256$.

(b) We expect it to be symmetric and bell-shaped and centered at the population proportion of 0.256.

B.15 (a) Both distributions are centered at the population parameter, so 0.05.

(b) The proportions for samples of size $n = 100$ go from about 0 to 0.12. The proportions for samples of size $n = 1000$ go from about 0.025 to 0.07.

(c) The standard error for samples of size $n = 100$ is about 0.02 (since it appears that about 95% of the data are between 0.01 and 0.09.) The standard error for samples of size $n = 1000$ is about 0.005 (since it appears that about 95% of the data are between 0.04 and 0.06.)

(d) A sample proportion of 0.08 is relatively likely from a sample of 100, but extremely unlikely with a sample size of 1,000.

B.17 (a) Answers will vary. Here is one possible set of randomly selected *Points* values.
Points: 7, 3, 19, 3, 48 $\bar{x} = 16.0$

(b) Answers will vary. Here is another possible set of randomly selected *Points* values.
Points: 64, 1, 36, 19, 27 $\bar{x} = 29.4$

(c) The mean number of points for all 24 players is $\mu = 25.67$ points for the season. Most sample means found in parts (a) and (b) will be somewhat close to this but not exactly the same.

(d) The distribution will be roughly symmetric with a peak at the center of 25.67. See the figure.

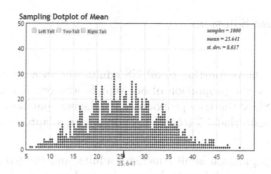

B.19 Answers will vary, but a typical distribution is shown below. The smallest mean is just above 5 and the largest is around 50 (but answers will vary). The standard deviation of these 1000 sample means is about 8.0.

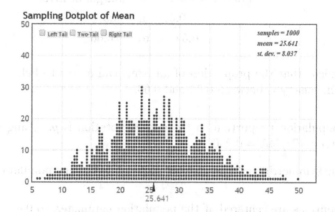

B.21 The p-value 0.0004 goes with the experiment showing significantly lower performance on material presented while the phone was ringing. The p-value 0.93 goes with the experiment measuring the impact of proximity of the student to the ringing phone. The p-value of 0.0004 shows very strong evidence that a ringing cell phone in class affects student learning.

B.23 This is a test for a difference in proportions, and we define p_F and p_N to be the proportion of men copying their partners sentence structure with a fertile partner and non-fertile partner, respectively. The hypotheses are:

$$H_0: \quad p_F = p_N$$
$$H_a: \quad p_F < p_N$$

The sample statistic is $\hat{p}_F - \hat{p}_N = 30/62 - 38/61 = 0.484 - 0.623 = -0.139$. In a randomization distribution such as the one below, we see that the p-value in the left tail beyond this point is 0.089. We do not reject H_0 at a 5% level, and do not find evidence that men's speech is affected by ovulating women. The results are borderline, though, and are significant at a 10% level. It might be worth continuing the experiment with a larger sample size.

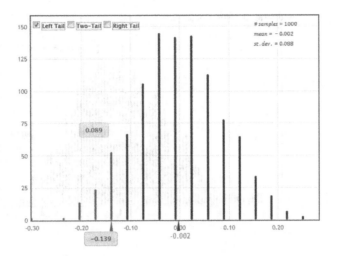

B.25 (a) Answers vary. One possible sample is: 120, 130, 150, 180, 120, 140, 200, 180, 170, 180. All values well above $100.

(b) Answers vary. One possible sample is: 70, 120, 90, 110, 80, 60, 110, 100, 80, 120. In this case, the sample mean is $\overline{x} = 94$ which is less than 100 so provides no evidence at all that the mean is larger than 100.

(c) Answers vary. One possible sample is: 90, 100, 70, 110, 120, 80, 140, 100, 80, 120. In this case, the sample mean is $\overline{x} = 101$ which is just barely bigger than $100. Since the sample mean is larger than $100, we have some evidence that the population mean will be larger than $100 but it is very weak evidence.

B.27 (a) The mean is $\overline{x} = 67.59$ and the standard deviation is $s = 50.02$.

(b) Select 20 values at random (with replacement) from the original set of skateboard prices and record the mean for those 20 values as the bootstrap statistic.

(c) We expect the bootstrap distribution to be symmetric and bell-shaped and to be centered at the sample mean: 67.59.

(d) We find the 95% confidence interval:

$$
\begin{aligned}
\overline{x} \quad &\pm \quad 2 \cdot SE \\
67.59 \quad &\pm \quad 2(10.9) \\
67.59 \quad &\pm \quad 21.8 \\
45.79 \quad &\text{to} \quad 89.39.
\end{aligned}
$$

We are 95% confident that the mean price of skateboards for sale online is between $45.79 and $89.39.

B.29 (a) For one set of 3000 bootstrap sample standard deviations shown below, the 2.5%-tile and 97.5%-tile are 15.1 and 35.5, respectively. Thus we can say with 95% confidence that the standard deviation of the number of penalty minutes awarded to all NHL players in a season is between 15.1 and 35 minutes.

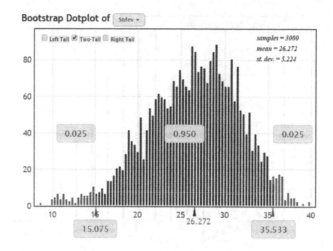

(b) The midpoint of the interval in part (a) is $(15.1 + 35.5)/2 = 25.3$ which is less than the standard deviation of the original sample, $s = 27.3$. In general, an interval based on bootstrap percentiles does not need to be centered at the original sample statistic.

B.31 (a) We compute the regression line to be

$$\widehat{PctRural} = 30.35 + 0.075 \cdot Area.$$

The slope of the line for this sample is 0.075.

(b) Using technology to produce the bootstrap distribution below for the sample slopes, we get a 95% confidence interval for the slope from -0.006 to 0.142. Answers will vary – for this small a sample with strongly skewed data the bootstrap slopes might contain some very extreme values. We are 95% confident that the slope of the regression line for all countries to predict percent rural from land area is between -0.006 and 0.142.

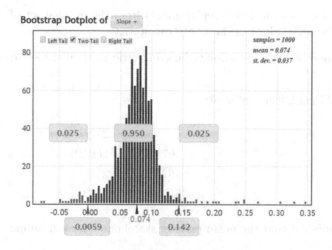

(c) The 95% confidence interval from part (b) is (-0.006, 0.142), so we barely capture the true population slope of 0. The lower bound is very close to zero, so this answer may vary, depending on the results of the simulation from part (b).

B.33 (a) Since we are looking at whether smoking has a negative effect, this is a one-tailed test.

(b) The null hypothesis will be that the two proportions are identical $H_0 : p_s = p_{ns}$, and the alternative is that the proportion of successful pregnancies will be less in the smoking group $H_a : p_s < p_{ns}$.

(c) We want the number assigned to each group to match the numbers in the original sample, so 135 women to the smoking group and 543 to the non-smoking. (Both of these values can be found in the two-way table.)

(d) In the original sample, there were 38 successful pregnancies in the 135 women in the smoking group. From the randomization distribution, it appears that about 40 of the 1000 values fall less than or equal to the count of 38 from the original sample, so the best estimate for the p-value is about $40/1000 = 0.04$.

B.35 (a) Sample A. The sample mean in A is around 43, while the sample mean in B is around 47. Sample sizes and variability are similar for both samples.

(b) Sample B. Both samples appear to have a mean near 46, but the variability is smaller in sample B so we can be more sure the mean is below 50. Also, sample B has few values above 50, while sample A has at least 25% of its values above 50 (since $Q_3 > 50$).

(c) Sample A. Both samples appear to have about the same mean and median (near 45) and similar variability, but sample A is based on a much larger sample size, so it would be more unusual to see that many values below 50 if $H_0 : \mu = 50$ were true.

B.37 (a) This is a population proportion so the correct notation is p. We have $p = 166/691 = 0.240$.

(b) Using technology we produce a sampling distribution (shown below) of 5000 sample proportions when samples of size $n = 40$ are drawn from a population with $p = 0.240$. We see that the distribution is relatively symmetric, bell-shaped and centered at the population proportion of 0.240, as we expect. We also see in the figure that the estimated standard error based on these 1000 simulated proportions is 0.068.

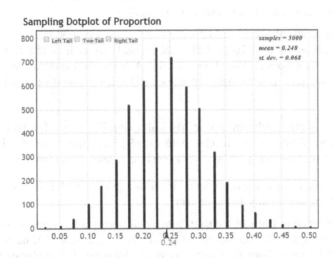

B.39 (a) Approximately the same. We expect both distributions to be approximately symmetric and bell-shaped.

(b) Different. The sampling distribution is centered at the value of the population parameter, while the bootstrap distribution is centered at the value of the sample statistic.

(c) Approximately the same. The standard error from the bootstrap distribution gives a good approximation to the standard error for the sampling distribution.

(d) Different. One value in the sampling distribution represents the statistic from a sample taken (without replacement) from the entire population, while one value in the bootstrap distribution represents the statistic from a sample taken with replacement from the original sample. In both cases, however, we compute the same statistic (mean, proportion, or whatever) and use the same sample size.

(e) Different. In order to create a sampling distribution, we need to know the data values for the entire population! In order to create a bootstrap distribution, we only need to know the values in one sample. This is what makes the bootstrap method so powerful.

B.41   (a) The hypotheses are $H_0 : p = 0.5$ vs $H_a : p > 0.5$, where p is the proportion of all games Paul the Octopus picks correctly.

(b) Answers vary, but 8 out of 8 heads should rarely occur.

(c) The proportion of heads in flipping a coin is $p = 0.5$ which matches the null hypothesis.

B.43 Since you are testing whether the coin is biased, the null hypothesis is $H_0 : p = 0.5$, where p is the proportion of heads for all flips. We assume the coin is fair and H_0 is true when we create the randomization distribution, so we expect the simulated sample $\hat{p}$ values to be centered around 0.5.

B.45   (a) The population is all American adults. The sample is the 7,293 people who were contacted.

(b) We are 95% sure that the proportion of all American adults planning to watch the game was between 57.3% and 59.7%.

(c) The center of the confidence interval is $(0.573+0.597)/2 = 0.585$, so we expect the estimated proportion to watch Super Bowl 50 is $\hat{p} = 0.586$, with a margin of error of 0.012.

B.47   (a) The hypotheses are $H_0 : \mu = 160$ vs $H_a : \mu < 160$, where μ is the mean score on the forced swim test for depressed mice after treatment with ketamine.

(b) The mean for the original sample of 10 mice is $\bar{x} = 135$ seconds and we need to match the null mean of $\mu = 160$ seconds so we add $160 - 135 = 25$ seconds to each of the ten data values and write each new score on a slip of paper. We choose a slip of paper at random (with replacement), write down the value, and continue until we have 10 values. The randomization statistic is the mean of those 10 values.

B.49 When the results of the study are not statistically significant, we fail to reject the null hypothesis (in this case that heavy cell phone use is unrelated to developing bring cancer). But that does not mean we "accept" that H_0 *must* be true, we just lack sufficiently convincing evidence to refute it.

(a) By not rejecting H_0 we are saying it is a plausible option, so it might be true that heavy cell phone use has no effect on developing brain cancer.

(b) There is some evidence in the sample that heavy cell phone users have a higher risk of developing brain cancer, just not enough to be considered statistically significant. Failing to reject H_0 means either H_0 or H_a might still be true. Note that any confidence interval that includes zero, contains both positive and negative values as plausible options. Hence the authors tell us that the question "remains open".

(c) We note that this study was an observational study and not an experiment, and thus, even if the results had been statistically significant, we would not make a cause/effect conclusion about this relationship. However, a *lack* of significant results does not rule this out as a plausible option.

B.51 (a) The parameter of interest is ρ, the correlation between score on the mouse grimace scale and pain intensity and duration. Since the study is investigating a positive relationship between these variables, the hypotheses are $H_0 : \rho = 0$ vs $H_a : \rho > 0$.

(b) Yes. If they conclude there is some relationship, the sample correlation must have been statistically significant (and positive).

(c) No. If the original correlation is statistically significant, a sample produced under the null hypothesis of no relationship should rarely give a correlation more extreme than was originally observed. This is what we mean by statistically significant.

(d) If the results of the original study were not significant, it would not be very unusual to get a sample correlation that extreme when H_0 is true. So it would not be very surprising to see a placebo give a larger correlaion.

B.53 (a) This is an upper tail test, so the p-value is the proportion of randomization samples with differences more than the observed $D = 0.79$. There are 27 dots to the right of 0.79 in the plot, so the p-value is $27/1000 = 0.027$.

(b) The randomization distribution depends only on H_0 so it would not change for $H_a : \mu_s \neq \mu_n$. For a two-tailed alternative we need to double the proportion in one tail, so the p-value is $2(0.027) = 0.054$.

B.55 (a) The sample proportion is 0.57 and this is the best point estimate we have of the population proportion of inaccurate classifications of truthful answers when under stress.

(b) Using the sample proportion and the bootstrap standard error we get an interval estimate for the true proportion using

$$
\begin{array}{ccc}
\hat{p} & \pm & 2 \cdot SE \\
0.57 & \pm & 2 \cdot 0.07 \\
0.57 & \pm & 0.14 \\
0.43 & \text{to} & 0.71
\end{array}
$$

(c) The proportion of false reports of lying is extremely large throughout our interval! So results from this lie detector should not hold up in court.

B.57 (a) Reject H_0. The mean resting metabolic rate is higher in lemurs after treatment with a resveratrol supplement.

(b) Reject H_0. The mean body mass gain is lower in lemurs after treatment with a resveratrol supplement.

(c) Reject H_0. The mean food intake in lemurs changes after treatment with a resveratrol supplement. We can't tell from the information given which way it changes.

(d) Do not reject H_0. There is not evidence of a change in mean locomotor activity in lemurs after treatment with a resveratrol supplement.

(e) Strongest evidence (smallest p-value) is in the test for mean body mass gain (p-value=0.007). Weakest evidence (largest p-value) is in the test for locomotor activity (p-value=0.980).

(f) Parts (b) and (d) remain the same with a 1% significance level, but parts (a) and (c) would change to "do not reject H_0" since those p-values are larger than 0.01. Thus the difference in mean metabolic rate and mean food intake would not be significant at a 1% level.

(g) We have strong evidence that the mean metabolic rate is lower when treating lemurs with resveratrol, very strong evidence that the mean body mass gain is lower, strong evidence that the mean food intake is different, but no evidence that mean locomotor activity is affected.

(h) If the lemurs represent a random sample of all lemurs, then we can generalize the findings based on this small p-value to conclude that the mean body mass gain is lower after four weeks of resveratrol supplements.

B.59 Using *StatKey* or other technology we create a bootstrap distribution using the original sample with a proportion of 28 autism cases out of 92 siblings ($\hat{p} = 0.304$). We find that a 99% confidence interval from this distribution goes from about 0.185 to 0.424. We are 99% sure that the percentage of siblings of children with autism likely to themselves have autism is between 18.5% and 42.4%.

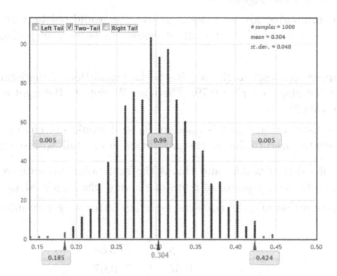

B.61 Using *StatKey* or other technology we construct a bootstrap distribution based on the mean commute distances for 5000 samples of size $n = 500$ taken (with replacement) from the original **CommuteAtlanta** distances. For the distribution shown below, the 5%-tile is 17.15 and the 95%-tile is 19.15 so the 90% confidence interval is (17.15, 19.15). We are 90% sure that the average distance to work for all commuters in metropolitan Atlanta is between 17.15 miles and 19.15 miles.

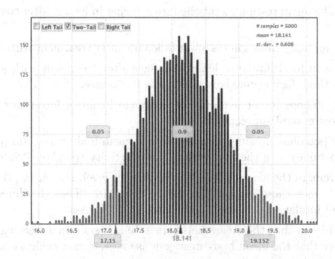

B.63 (a) Select a sample of size 500, with replacement, from the original **CommuteAtlanta** data and compute the correlation between *Distance* and *Time* for that sample.

(b) Among one set of 1000 bootstrap correlations show below, the 0.5%-tile is 0.706 and the 99.5%-tile is 0.876. The 99% confidence interval is (0.706,0.876). We are 99% sure that the correlation between distance and time of Atlanta commutes is somewhere between 0.706 and 0.876.

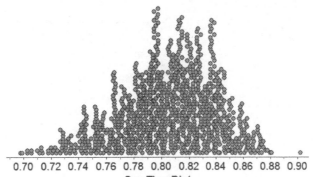

(c) For this set of bootstrap correlations the 95% confidence interval goes from 0.729 to 0.867. The 90% confidence interval goes from 0.742 to 0.859.

(d) As the confidence level decreases from 99% to 95% to 90% the confidence intervals get narrower.

B.65 (a) For the small p-value of 0.0012, we expect about $0.0012 \cdot 1000 = 1.2$ or about one time out of every 1000 to be as extreme as the difference observed, if the questions are equally difficult.

(b) The p-value is very small, so seeing this large a difference would be very unusual if the two questions really were equally difficult. Thus we conclude that there is a difference in the average difficulty of the two questions.

(c) There is nothing in the information given that indicates which question had the higher mean, so we can't tell which of the two questions is the easier one.

B.67 (a) Let ρ denote the correlation between pH and mercury in all Florida lakes. The question of interest is whether or not this correlation is negative, so we use a one-tailed test, with hypotheses $H_0 : \rho = 0$ vs $H_a : \rho < 0$.

(b) We want to find a point that has roughly 30% of the randomization distribution in the lower tail below it. This should occur somewhere between $r = -0.20$ and $r = -0.10$, perhaps $r \approx = -0.15$. (It is difficult to determine this point very precisely from the plot.)

(c) We want to find a point that has only about 1% of the randomization distribution in the lower tail below it. This should occur around $r \approx -0.50$.

B.69 (a) The hypotheses are $H_0 : p_1 = p_2$ vs $H_a : p_1 > p_2$, where p_1 and p_2 are the proportion with reduced pain when using cannabis and a placebo, respectively.

(b) The sample statistics are $\hat{p}_1 = 14/27 = 0.519$ and $\hat{p}_2 = 7/28 = 0.250$. Since $\hat{p}_1 > \hat{p}_2$, the sample statistics are in the direction of H_a.

(c) If the FDA requires very strong evidence to reject H_0, they should choose a small significance level, such as $\alpha = 0.01$.

(d) In this situation, under a null hypothesis that $H_0 : p_1 = p_2$, pain response would be the same whether in the cannabis or placebo group. The randomization distribution for $\hat{p}_1 - \hat{p}_2$ should be centered at zero.

(e) We draw a bell-shaped curve, centered at 0, and roughly locate the original sample statistic, $\hat{p}_1 - \hat{p}_2 = 0.519 - 0.250 = 0.269$, so that the area in the right tail is only about 0.02.

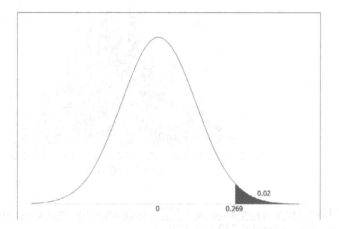

(f) A p-value as small as 0.02 gives fairly strong evidence to reject $H_0 : p_1 = p_2$.

(g) If the FDA uses a small $\alpha = 0.01$, the p-value is not less than α, so we do not reject H_0. Although the sample results are suggestive of a benefit to using cannabis for pain reduction, they are not sufficiently strong to conclude (at a 1% significance level) that the proportion of patients having reduced pain after using cannabis is more than the proportion who are helped by a placebo.

B.71   (a) It doesn't matter whether we test the proportion who plan to vote for Obama or the proportion who plan to vote for McCain, since if we know one we know the other. (For example, if 54% plan to vote for Obama, then the other 46% plan to vote for McCain and vice versa.) If we let p_O represent the proportion that plan to vote for Obama, then we are testing to see if the proportion is greater than 0.5 so the hypotheses are

$$H_0 : \quad p_O = 0.5$$
$$H_a : \quad p_O > 0.5$$

Notice that if we use p_M as the proportion that plan to vote for McCain, the alternative hypothesis would be $p_M < 0.5$. The two ways of doing the test are equivalent.

(b) We assume from the null hypothesis that half ($p_O = 0.5$) plan to vote for Obama and half for McCain, so we might flip a coin to determine the outcome for each simulated voter. Since the sample size was 1057, to create a simulated sample, we need to flip a coin 1057 times (or flip 1057 coins!) and record the number of heads (Obama) and tails (McCain) in the 1057 flips. We record the proportion for Obama as one randomization statistic. We do this many times to collect 1000 or more such randomization statistics and those values form our randomization distribution. There are many other ways to create the randomization distribution, but every such method should use the null assumption of half supporting each candidate and every method should use the sample size of 1057.

B.73　(a) In a Type I error, we conclude that treated wipes prevent infection, when actually they don't.

(b) In a Type II error, we conclude that treated wipes are not shown to be effective, when actually they help prevent infections.

(c) A smaller significance level means we need more evidence to reject H_0. We would want a smaller significance level in the second situation (harmful side effects) so it has to be very clear that the treated wipes help prevent infection, since we don't want to put people at risk for side effects if the benefit isn't definite.

(d) The p-value (0.32) is not small, so we do not reject H_0. The study does not provide sufficient evidence to show that treated wipes are more effective at reducing the proportion of infected babies than sterile wipes.

(e) Not necessarily. The results of the test are inconclusive when the p-value is not small. Either H_0 or H_a could still be valid, so the treated wipes might help prevent infections and the study just didn't accumulate enough evidence to verify it.

B.75 We use technology to create a randomization distribution (such as shown below) and then check to see how extreme the original sample statistic of $r = -0.575$ is in that distribution. None of the randomization statistics in this distribution are beyond (less than since this is a lower-tail test) the observed $r = -0.575$. Thus the p-value ≈ 0.000. Based on this very small p-value, we have strong evidence to reject H_0 and conclude that there is negative correlation between pH levels and the amount of mercury in fish of Florida lakes. More acidic lakes tend to have higher mercury levels in the fish.

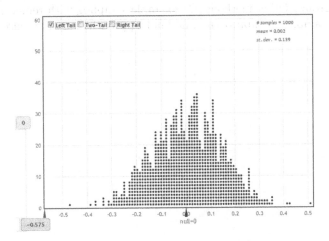

B.77 For one set of 1000 randomization samples (shown below), 242 of the sample correlations are more than the observed $r = 0.279$ which gives a p-value of 0.242. This p-value is not smaller than any reasonable significance level, so we do not have sufficient evidence to reject H_0. Based on this sample of 8 patients, we can not conclude that there must be a positive association between heart rate and systolic blood pressure for 55 year-old patients at this ICU.

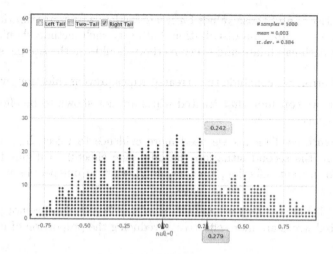

B.79 The null hypothesis for this correlation test is that the correlation is zero. To create the randomization samples, we match the null hypothesis. In this situation, that means height and salary are completely unrelated. We might randomly scramble the height values and assign them to the original subjects/salaries. Compute the correlation, r, between those random heights and the actual salaries.

B.81 The null hypothesis for this difference in means test is that the means of the two groups are the same. To create the randomization samples, we match the null hypothesis. In this situation, that means whether or not a customer is approached has no effect on sales. We might randomly scramble the labels for type of store ("approach" or "not approach") and assign them to the actual sales values. Compute the difference in the mean sales, $\bar{x}_a - \bar{x}_{na}$, between the stores assigned to the "approach" group and those randomly put in the "not approach" group. Other methods, for example sampling with replacement from the pooled sales values to simulate new samples of "approach" and "not approach" sales, are also acceptable.

Section 5.1 Solutions

5.1 The area in the right tail more extreme than $z = 2.20$ is 0.014.

5.3 The area in the right tail more extreme than $z = -1.25$ is 0.894.

5.5 The area in the left tail more extreme than $z = -1.75$ is 0.040.

5.7 For this test of a mean we compare the sample mean to the hypothesized mean and divide by the standard error.

$$z = \frac{\text{Sample mean} - \text{Null mean}}{SE} = \frac{82.4 - 80}{0.8} = 3.0$$

5.9 For this test of a proportion we compare the sample proportion to the hypothesized proportion and divide by the standard error.

$$z = \frac{\text{Sample proportion} - \text{Null proportion}}{SE} = \frac{0.41 - 0.5}{0.07} = -1.29$$

5.11 The relevant statistic here is the difference in proportions. From the null hypothesis, we see that $p_1 - p_2 = 0$. We have:

$$z = \frac{\text{Sample difference in proportions} - \text{Null difference in proportions}}{SE} = \frac{(0.18 - 0.23) - 0}{0.05} = -1.0$$

5.13   (a) Using technology the standard normal area above $z = 0.84$ is 0.20. The p-value for an upper tail test is 0.20.

(b) Using technology the standard normal area below $z = -2.38$ is 0.0087. The p-value for a lower tail test is 0.0087.

(c) Using technology the standard normal area above $z = 2.25$ is 0.012. Double this to find the p-value for a two-tail test, $2 \cdot 0.0122 = 0.024$.

These three p-values are shown as areas below.

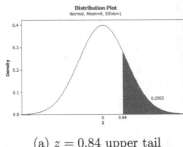

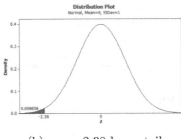

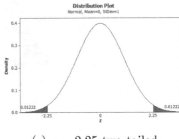

(a) $z = 0.84$ upper tail   (b) $z = -2.38$ lower tail   (c) $z = 2.25$ two-tailed

5.15   (a) The plot on the left shows the area above 65 on a $N(60, 10)$ distribution to be 0.309. We could also standardize the endpoint with

$$z = \frac{65 - 60}{10} = 0.5$$

and use the $N(0,1)$ curve on the right below to find the area.

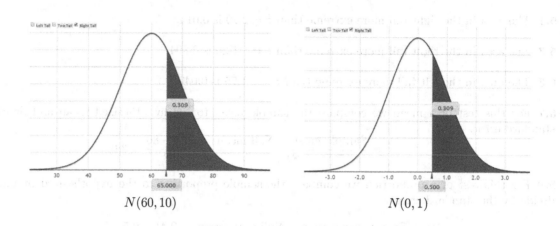

$$N(60,10) \qquad\qquad\qquad N(0,1)$$

(b) The plot on the left shows the area below 48 on a $N(60,10)$ distribution to be 0.115. We could also standardize the endpoint with

$$z = \frac{48 - 60}{10} = -1.2$$

and use the $N(0,1)$ curve on the right below to find the area.

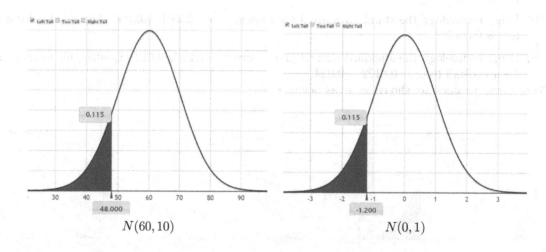

$$N(60,10) \qquad\qquad\qquad N(0,1)$$

5.17   (a) The plot on the left shows the area above 140 on a $N(160,25)$ distribution to be 0.788. We could also standardize the endpoint with

$$z = \frac{140 - 160}{25} = -0.8$$

and use the $N(0,1)$ curve on the right below to find the area.

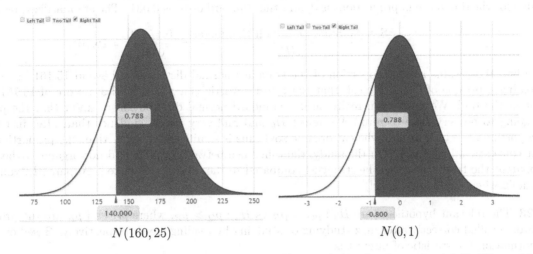

$$N(160, 25) \qquad\qquad N(0,1)$$

(b) The plot on the left shows the area below 200 on a $N(160, 25)$ distribution to be 0.945. We could also standardize the endpoint with

$$z = \frac{200 - 160}{25} = 1.6$$

and use the $N(0,1)$ curve on the right below to find the area.

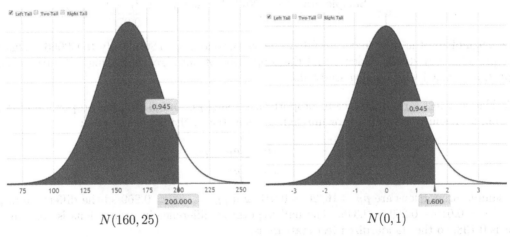

$$N(160, 25) \qquad\qquad N(0,1)$$

5.19 The sample statistic is the difference is proportions, which is $\hat{p}_o - \hat{p}_c = 0.42 - 0.40 = 0.02$. The null value for the difference in proportions is 0, and the standard error is 0.031. The standardized test statistic is

$$z = \frac{\text{Sample Statistic} - \text{Null Parameter}}{SE} = \frac{0.02 - 0}{0.031} = 0.645.$$

We find the p-value as the proportion of the standard normal distribution beyond 0.645 in the right tail, yielding a p-value of 0.259. This p-value is not significant at a 5% level, so we do not reject H_0. After 25 days, we don't see evidence of a significant difference in the proportion alive between those eating organic bananas and those eating conventional bananas. (In fact, throughout the study, it didn't seem to matter whether bananas were organic or not. With the other types of food tested, the difference became significant after enough time passed.)

5.21 The sample statistic is the difference is proportions, which is $\hat{p}_o - \hat{p}_c = 0.79 - 0.32 = 0.47$. The null value for the difference in proportions is 0, and the standard error is 0.031. The standardized test statistic is

$$z = \frac{\text{Sample Statistic} - \text{Null Parameter}}{SE} = \frac{0.47 - 0}{0.031} = 15.161.$$

We find the p-value as the proportion of the standard normal distribution beyond 15.161 in the right tail, yielding a p-value of 0.000. (Recall that the z test statistic is a z-score, and a z-score of 15.161 is very far out in the tail! We don't even really need a standard normal distribution to know that the p-value here is going to be very close to zero.) We reject H_0 and find very strong evidence that, after just 8 days, the proportion alive of fruit flies who eat organic soybeans is significantly higher than the proportion alive who eat conventional soybeans. (In the study, the difference between organic and not-organic soybeans became apparent the fastest and was the strongest compared to the other foods tested. Apparently, you should get organic when buying soybeans!)

5.23 The relevant hypotheses are $H_0 : p_Q = p_R$ vs $H_a : p_Q > p_R$, where p_Q and p_R are the proportions of words recalled correctly after quiz studying or studying by reading alone, respectively. Based on the sample information the statistic of interest is

$$\hat{p}_Q - \hat{p}_R = 0.42 - 0.15 = 0.27$$

The standard error of this statistic is given as $SE = 0.07$ and the null hypothesis is that the difference in the proportions for the two group is zero. We compute the standardized test statistic with

$$z = \frac{\text{Sample Statistic} - \text{Null Parameter}}{SE} = \frac{0.27 - 0}{0.07} = 3.86$$

Using technology, the area under a $N(0,1)$ curve beyond $z = 3.86$ is only 0.000056. This very small p-value provides very strong evidence that the proportion of words recalled using self-quizzes is more than the proportion recalled with reading study alone.

5.25 This is a test for a difference in proportions. Using p_R and p_U for the proportion of dogs to follow a cue from a person who is reliable or unreliable, respectively, the hypotheses are:

$$H_0 : \quad p_R = p_U$$
$$H_a : \quad p_R \neq p_U$$

The sample proportions are $\hat{p}_R = 16/26 = 0.615$ and $\hat{p}_U = 7/26 = 0.269$ so the difference in proportions is $\hat{p}_R - \hat{p}_U = 0.615 - 0.269 = 0.346$. The null hypothesis difference in proportions is zero and the standard error is 0.138, so the standardized test statistic is:

$$z = \frac{\text{Sample Statistic} - \text{Null Parameter}}{SE} = \frac{0.346 - 0}{0.138} = 2.507$$

The p-value is the area more extreme than this test statistic. Using technology, we see that the area beyond this value in the right tail of a standard normal distribution is 0.006. This is a two-tail test, so we have

$$\text{p-value} = 2(0.006) = 0.012.$$

At a 5% level, we reject H_0 and find evidence that the proportion of dogs following a pointing cue is different depending on whether the previous cue was reliable or not. The proportion that follow the cue appears to be higher when the person is reliable.

5.27 Under the null hypothesis the randomization distribution should be centered at zero with an estimated standard error of 0.393. The standardized test statistic is

$$z = \frac{0.79 - 0}{0.393} = 2.01$$

Since this is an upper tail test, the p-value is the area in a standard normal distribution above 2.01. Technology or a table shows this area to be 0.022 which is a fairly small p-value (significant at a 5% level). This gives support for the alternative hypothesis that students who smile during the hearing will, on average, tend to get more leniency than students with a neutral expression.

5.29 This is a test for a single mean. Using μ to represent the mean number of days meeting the goal, for people in a 100-day program to encourage exercise. The hypotheses are:

$$H_0: \quad \mu = 35$$
$$H_a: \quad \mu > 35$$

The sample statistic is $\overline{x} = 36.5$, the null value for the mean is 35, and the standard error for the estimate is given as 1.80. The standardized test statistic is:

$$z = \frac{\text{Sample Statistic} - \text{Null Parameter}}{SE} = \frac{36.5 - 35}{1.80} = 0.833$$

The p-value is the area more extreme than this test statistic. Using technology, we see that the area beyond this value in the right tail of a standard normal distribution is 0.202. We have

$$\text{p-value} = 0.202.$$

At a 5% level, we do not reject H_0. We do not find enough evidence to conclude that the mean number of days meeting the goal will be greater than 35.

5.31 (a) This is a test for a difference in proportions. Using p_G and p_I for the proportion quitting smoking for those in a group program or those in an individual program, respectively, the hypotheses are:

$$H_0: \quad p_G = p_I$$
$$H_a: \quad p_G \neq p_I$$

We see that $\hat{p}_G = 148/1080 = 0.137$ and $\hat{p}_I = 120/990 = 0.121$. The sample statistic is $\hat{p}_G - \hat{p}_I = 0.137 - 0.121 = 0.016$.

(b) A randomization distribution for the difference in proportion for 3000 randomizations is shown below. The proportion of samples more extreme than the difference of 0.016 in the original sample gives a p-value of $2 \cdot 0.143 = 0.286$.

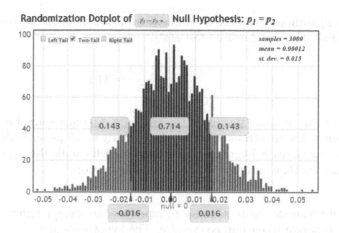

(c) The standard error in the randomization distribution above is $SE = 0.015$ and the null hypothesis is that the difference $p_G - p_I = 0$, so we use a $N(0, 0.015)$ distribution to find the p-value. The area above the original difference of 0.016 in the figure on the left below is 0.143. This gives a p-value of $2 \cdot 0.143 = 0.286$.

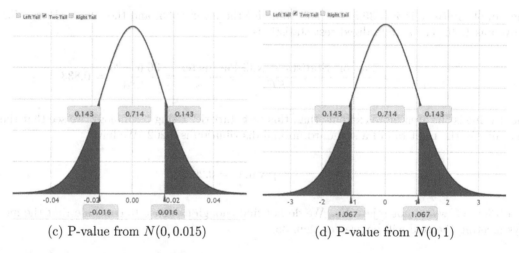

(c) P-value from $N(0, 0.015)$ (d) P-value from $N(0, 1)$

(d) The null hypothesis difference in proportions is zero and the standard error from the randomization distribution in (b) is 0.015, so the standardized test statistic is:

$$z = \frac{\text{Sample Statistic} - \text{Null Parameter}}{SE} = \frac{0.016 - 0}{0.015} = 1.067$$

The p-value is the area more extreme than this test statistic. Using technology (as in the figure on the right above), we see that the area beyond this value in the right tail of a standard normal distribution is 0.143. This is a two-tail test, so we have

$$\text{p-value} = 2(0.143) = 0.286.$$

(e) The p-values are the same (although a different set of randomizations might give a slightly different p-value). In every case, at a 5% level, we do not reject H_0. We do not find enough evidence to conclude that it makes a difference whether a smoker is in a group program or an individual program.

5.33   (a) The sample statistic is $\hat{p} = 4/20 = 0.20$, the proportion under the null hypothesis is 0.10, and the standard error of proportions based on samples of size $n = 20$ (from the StatKey figure) is SE=0.067. The standardized test statistic is
$$z = \frac{0.20 - 0.1}{0.067} = 1.49$$

   (b) Using technology, the area above 1.49 for a standard normal density is 0.068.

   (c) The p-value in the randomization distribution (0.137) is quite a bit larger than the p-value based on the normal distribution (0.068). This is a fairly small sample and the randomization distribution is not symmetric, so the p-value based on a normal distribution is not accurate. Fortunately, the randomization procedure gives a reasonable way to estimate the p-value in cases where the normal distribution is not appropriate.

Section 5.2 Solutions

5.35 (a) For 86% confidence we use technology to find the middle 92% of a standard normal distribution (leaving 7% in each tail) to give $z^* = 1.476$.

(b) For 94% confidence we use technology to find the middle 92% of a standard normal distribution (leaving 3% in each tail) to give $z^* = 1.881$.

(c) For 96% confidence we use technology to find the middle 92% of a standard normal distribution (leaving 2% in each tail) to give $z^* = 2.054$.

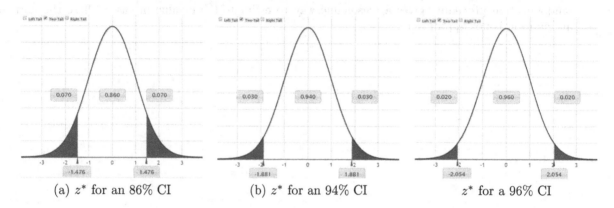

(a) z^* for an 86% CI   (b) z^* for an 94% CI z^* for a 96% CI

5.37 For a 95% confidence interval, we have $z^* = 1.96$. The confidence interval for μ is:

$$
\begin{array}{rcl}
\text{Sample statistic} & \pm & z^* \cdot SE \\
\overline{x} & \pm & z^* \cdot SE \\
72 & \pm & 1.96(1.70) \\
72 & \pm & 3.332 \\
68.668 & \text{to} & 75.332.
\end{array}
$$

5.39 For a 99% confidence interval, we have $z^* = 2.576$. The confidence interval for p is:

$$
\begin{array}{rcl}
\text{Sample statistic} & \pm & z^* \cdot SE \\
\hat{p} & \pm & z^* \cdot SE \\
0.78 & \pm & 2.576(0.03) \\
0.78 & \pm & 0.077 \\
0.703 & \text{to} & 0.857.
\end{array}
$$

5.41 For a 95% confidence interval, we have $z^* = 1.96$. The confidence interval for $\mu_1 - \mu_2$ is:

$$
\begin{array}{rcl}
\text{Sample statistic} & \pm & z^* \cdot SE \\
(\overline{x}_1 - \overline{x}_2) & \pm & z^* \cdot SE \\
(256 - 242) & \pm & 1.96(6.70) \\
14 & \pm & 13.132 \\
0.868 & \text{to} & 27.132.
\end{array}
$$

5.43 To find a confidence interval using the normal distribution, we use

$$\text{Sample statistic } \pm z^* \cdot SE.$$

We are finding a confidence interval for a proportion, so the statistic from the original sample is $\hat{p} = 0.195$. For a 90% confidence interval, we use $z^* = 1.645$ and we have $SE = 0.009$. Putting this information together, we have

$$
\begin{array}{ccl}
\hat{p} & \pm & z^* \cdot SE \\
0.195 & \pm & 1.645 \cdot (0.009) \\
0.195 & \pm & 0.0148 \\
0.180 & \text{to} & 0.210
\end{array}
$$

We are 90% sure that the percent of people aged 12 to 19 who have at least slight hearing loss is between 18% and 21%.

5.45 To find a confidence interval using the normal distribution, we use

$$\text{Sample statistic } \pm z^* \cdot SE.$$

We are finding a confidence interval for a mean, so the statistic from the original sample is $\overline{x} = 57.55$. For a 95% confidence interval, we use $z^* = 1.960$ and we have $SE = 1.42$. Putting this information together, we have

$$
\begin{array}{ccl}
\overline{x} & \pm & z^* \cdot SE \\
57.55 & \pm & 1.960 \cdot (1.42) \\
57.55 & \pm & 2.78 \\
54.77 & \text{to} & 60.33
\end{array}
$$

We are 95% confident that the average age of patients admitted to the intensive care unit at this hospital is between 54.8 years old and 60.3 years old.

5.47 To find a confidence interval using the normal distribution, we use

$$\text{Sample statistic } \pm z^* \cdot SE.$$

We are finding a confidence interval for a difference in proportions, so the relevant statistic from the original sample is $\hat{p}_Q - \hat{p}_R = 0.42 - 0.15 = 0.27$. For a 99% confidence interval, we use $z^* = 2.576$ and we have $SE = 0.07$. Putting this information together, we have

$$
\begin{array}{ccl}
(\hat{p}_Q - \hat{p}_R) & \pm & z^* \cdot SE \\
(0.42 - 0.15) & \pm & 2.576 \cdot (0.07) \\
0.27 & \pm & 0.18 \\
0.09 & \text{to} & 0.45
\end{array}
$$

We are 99% sure that the proportion of words remembered correctly will be between 0.09 and 0.45 higher for people who incorporate quizzes into their study habits.

5.49 To find a confidence interval using the normal distribution, we use

$$\text{Sample statistic} \pm z^* \cdot SE.$$

We are finding a confidence interval for a proportion, so the statistic from the original sample is $\hat{p} = 53/194 = 0.273$. For a 95% confidence interval, we use $z^* = 1.960$. We are given that $SE = 0.032$. Putting this information together, we have

$$
\begin{array}{ccl}
\hat{p} & \pm & z^* \cdot SE \\
0.273 & \pm & 1.960 \cdot (0.032) \\
0.273 & \pm & 0.063 \\
0.210 & \text{to} & 0.336
\end{array}
$$

We are 95% sure that the proportion of all US adults ages 18 to 24 who have used online dating is between 0.210 and 0.336.

5.51 (a) We see from the two-way table that the proportion of college graduates using online dating is $\hat{p}_C = 157/823 = 0.191$ and the proportion of high school graduates using online dating is $\hat{p}_H = 70/635 = 0.110$. The difference in proportions is $\hat{p}_C - \hat{p}_H = 0.191 - 0.110 = 0.081$.

  (b) To find a confidence interval using the normal distribution, we use

$$\text{Sample statistic} \pm z^* \cdot SE.$$

We are finding a confidence interval for a difference in proportions. For a 99% confidence interval, we use $z^* = 2.576$. We are given that $SE = 0.019$. Putting this information together, we have

$$
\begin{array}{ccl}
(\hat{p}_C - \hat{p}_H) & \pm & z^* \cdot SE \\
0.081 & \pm & 2.576 \cdot (0.019) \\
0.081 & \pm & 0.049 \\
0.032 & \text{to} & 0.130
\end{array}
$$

We are 99% sure that the proportion of all college graduates using online dating is between 0.032 and 0.130 higher than the proportion of all HS graduates using online dating.

  (c) No, it is not plausible that the proportions using online dating are the same for both groups since 0 (no difference) is not in the confidence interval for $p_C - p_H$.

5.53 For the original 25 Mustangs in the sample the mean price (in \$1000s) is $\overline{x} = 15.98$. We form a bootstrap distribution by sampling with replacement from the prices given in the original data set and finding the mean for each sample. One such set of 5000 bootstrap means is shown below.

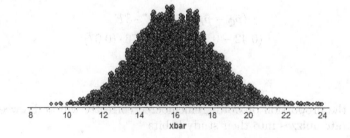

(a) For a 95% confidence interval we find the 2.5%-tile (11.88) and 97.5%-tile (20.44) from the distribution of bootstrap means. This gives a 95% confidence interval of (11.88, 20.44). Based on this analysis, we are 95% sure that the mean price of Mustangs for sale on the internet is between \$11,880 and \$20,440.

(b) The standard deviation of the 5000 bootstrap means is $SE = 2.202$. The distribution is reasonably normal so we use the $N(0,1)$ endpoints $z^* = \pm 1.96$ for a 95% confidence interval,

$$15.98 \pm 1.96 \cdot 2.202 = 15.98 \pm 4.316 = (11.664, 20.296)$$

Based on this analysis we are 95% sure that the mean price for Mustangs for sale on the Internet is between \$11,664 and \$20,296.

5.55 (a) The standard error for one set of 1000 bootstrap correlations is $SE = 0.0355$. Answers may vary slightly with other simulated bootstrap distributions.

(b) For a 90% confidence interval based on a normal distribution, the z^* value is 1.645. To find the confidence interval we use

$$0.807 \pm 1.645 \cdot 0.0355 = 0.807 \pm 0.058 = (0.749, 0.865)$$

We are 90% sure that the correlation between distance and time for all Atlanta commutes is somewhere between 0.749 and 0.865.

5.57 (a) For a 99% confidence interval the standard normal value leaving 0.5% in each tail is $z^* = 2.576$. From the bootstrap distribution we estimate the standard error of the correlations to be 0.205. Thus the 99% confidence interval based on a normal distribution would be

$$0.37 \pm 2.576 \cdot 0.205 = 0.37 \pm 0.528 = (-0.158, 0.898)$$

(b) The bootstrap distribution of correlations is somewhat right skewed, while the normal-based interval assumes the distribution is symmetric and bell-shaped.

Section 6.1-D Solutions

6.1 The sample proportions will have a standard error of

$$SE = \sqrt{\frac{p(1-p)}{n}} = \sqrt{\frac{0.25(1-0.25)}{50}} = 0.061$$

6.3 The sample proportions will have a standard error of

$$SE = \sqrt{\frac{p(1-p)}{n}} = \sqrt{\frac{0.90(1-0.90)}{60}} = 0.039$$

6.5 The sample proportions will have a standard error of

$$SE = \sqrt{\frac{p(1-p)}{n}} = \sqrt{\frac{.08(0.92)}{300}} = 0.016$$

6.7 We compute the standard errors using the formula:

$$n = 30: \quad SE = \sqrt{\frac{p(1-p)}{n}} = \sqrt{\frac{0.4(0.6)}{30}} = 0.089$$

$$n = 200: \quad SE = \sqrt{\frac{p(1-p)}{n}} = \sqrt{\frac{0.4(0.6)}{200}} = 0.035$$

$$n = 1000: \quad SE = \sqrt{\frac{p(1-p)}{n}} = \sqrt{\frac{0.4(0.6)}{1000}} = 0.015$$

We see that as the sample size goes up, the standard error goes down. If the standard error goes down, the sample proportions are less spread out from the population proportion, so the accuracy is better.

6.9 In each case, we determine whether $np \geq 10$ and $n(1-p) \geq 10$.

(a) Yes, the conditions apply, since $np = 500(0.1) = 50$ and $n(1-p) = 500(1-0.1) = 450$.

(b) Yes, the conditions apply, since $np = 25(0.5) = 12.5$ and $n(1-p) = 25(1-0.5) = 12.5$.

(c) No, the conditions do not apply, since $np = 30(0.2) = 6 < 10$.

(d) No, the conditions do not apply, since $np = 100(.92) = 92$ but $n(1-p) = 100(1-0.92) = 8 < 10$.

Section 6.1-CI Solutions

6.11 The sample size is large enough to use the normal distribution. For a confidence interval using the normal distribution, we use

$$\text{Sample statistic} \pm z^* \cdot SE.$$

The relevant sample statistic for a confidence interval for a proportion is $\hat{p} = 0.38$. For a 95% confidence interval, we have $z^* = 1.96$, and the standard error is $SE = \sqrt{\hat{p}(1-\hat{p})/n}$. The confidence interval is

$$\hat{p} \quad \pm \quad z^* \sqrt{\frac{\hat{p}(1-\hat{p})}{n}}$$
$$0.38 \quad \pm \quad 1.96 \cdot \sqrt{\frac{0.38(0.62)}{500}}$$
$$0.38 \quad \pm \quad 0.043$$
$$0.337 \quad \text{to} \quad 0.423$$

The best estimate for p is 0.38, the margin of error is ± 0.043, and the 95% confidence interval for p is 0.337 to 0.423.

6.13 The sample size is large enough to use the normal distribution, since there are 62 yes answers and 28 other answers, both bigger than 10. For a confidence interval using the normal distribution, we use

$$\text{Sample statistic} \pm z^* \cdot SE.$$

The relevant sample statistic for a confidence interval for a proportion is $\hat{p}$, and in this case we have $\hat{p} = 62/90 = 0.689$ with $n = 90$. For a 99% confidence interval, we have $z^* = 2.576$, and the standard error is $SE = \sqrt{\hat{p}(1-\hat{p})/n}$. The confidence interval is

$$\hat{p} \quad \pm \quad z^* \sqrt{\frac{\hat{p}(1-\hat{p})}{n}}$$
$$0.689 \quad \pm \quad 2.576 \cdot \sqrt{\frac{0.689(0.311)}{90}}$$
$$0.689 \quad \pm \quad 0.126$$
$$0.563 \quad \text{to} \quad 0.815$$

The best estimate for the proportion who will answer yes is 0.689, the margin of error is ± 0.126, and the 99% confidence interval for the proportion who will answer yes is 0.563 to 0.815. Note that the margin of error is quite large since the sample size is relatively small.

6.15 The desired margin of error is $ME = 0.05$ and we have $z^* = 1.96$ for 95% confidence. Since we are given no information about the population parameter, we use the conservative estimate $\tilde{p} = 0.5$. We use the formula to find sample size:

$$n = \left(\frac{z^*}{ME}\right)^2 \tilde{p}(1-\tilde{p}) = \left(\frac{1.96}{0.05}\right)^2 (0.5 \cdot 0.5) = 384.2.$$

We round up to $n = 385$. In order to ensure that the margin of error is within the desired $\pm 5\%$, we should use a sample size of 385 or higher.

6.17 The desired margin of error is $ME = 0.03$ and we have $z^* = 1.645$ for 90% confidence. We estimate that p is about 0.3, so we use $\tilde{p} = 0.3$. We use the formula to find sample size:

$$n = \left(\frac{z^*}{ME}\right)^2 \tilde{p}(1 - \tilde{p}) = \left(\frac{1.645}{0.03}\right)^2 (0.3 \cdot 0.7) = 631.4.$$

We round up to $n = 632$. In order to ensure that the margin of error is within the desired $\pm 3\%$, we should use a sample size of 632 or higher.

6.19   (a) The sample proportion is $\hat{p} = 0.35$, the sample size is $n = 140$, and for 95% confidence, we have $z^* = 1.96$. The 95% confidence interval is given by

$$
\begin{array}{ccc}
Statistic & \pm & z^* \cdot SE \\[6pt]
\hat{p} & \pm & z^* \cdot \sqrt{\dfrac{\hat{p}(1 - \hat{p})}{n}} \\[10pt]
0.35 & \pm & 1.96 \cdot \sqrt{\dfrac{0.35(0.65)}{140}} \\[10pt]
0.35 & \pm & 0.079 \\[4pt]
0.271 & \text{to} & 0.429.
\end{array}
$$

We are 95% sure that the proportion of US household cats that hunt outdoors is between 0.271 and 0.429.

   (b) We see that $p = 0.45$ is not in the confidence interval, so 0.45 is not a plausible value for the proportion. However, $p = 0.30$ is in the confidence interval, so 0.30 is a plausible value.

6.21 The sample size is definitely large enough to use the normal distribution. For a confidence interval using the normal distribution, we use

$$\text{Sample statistic} \pm z^* \cdot SE.$$

The relevant sample statistic for a confidence interval for a proportion is $\hat{p} = 0.20$. For a 99% confidence interval, we have $z^* = 2.576$, and the standard error is $SE = \sqrt{\hat{p}(1 - \hat{p})/n}$. The confidence interval is

$$
\begin{array}{ccc}
\hat{p} & \pm & z^* \sqrt{\dfrac{\hat{p}(1 - \hat{p})}{n}} \\[10pt]
0.20 & \pm & 2.576 \cdot \sqrt{\dfrac{0.20(0.80)}{1000}} \\[10pt]
0.20 & \pm & 0.033 \\[4pt]
0.167 & \text{to} & 0.233
\end{array}
$$

We are 99% confident that the proportion of US adults who say they never exercise is between 0.167 and 0.233. The margin of error is $\pm 3.3\%$.

6.23 The sample size is clearly large enough to use the formula based on the normal approximation, since there are well more than 10 responses in each category.

   (a) The proportion in the sample who disagreed is $\hat{p} = 1812/2625 = 0.69$ and $z^* = 1.645$ for 90% confidence, so we have

$$0.69 \pm 1.645 \sqrt{\frac{0.69(1 - 0.69)}{2625}} = 0.69 \pm 0.015 = (0.675, 0.705)$$

We are 95% sure that between 67.5% and 70.5% of people would disagree with the statement "There is only one true love for each person."

(b) The proportion in the sample who answered "don't know" is $\hat{p} = 78/2625 = 0.03$ so the 90% confidence interval is

$$0.03 \pm 1.645 \sqrt{\frac{0.03(1 - 0.03)}{2625}} = 0.03 \pm 0.005 = (0.025, 0.035)$$

We are 90% sure that between 2.5% and 3.5% of people would respond with "don't know."

(c) The estimated proportion of people who disagree (which is closer to 0.5) has a larger margin of error.

6.25 (a) In each case, we use

$$\hat{p} \pm z^* \sqrt{\frac{\hat{p}(1 - \hat{p})}{n}}$$

for the confidence interval, using $z^* = 1.96$ for a 95% confidence interval. In every case, $n = 970$.

- For percent updating their status, we have:

$$0.15 \pm 1.96 \cdot \sqrt{\frac{0.15(0.85)}{970}} = 0.15 \pm 0.022 = \text{the interval from } 0.128 \text{ to } 0.172.$$

  We are 95% confident that between 12.8% and 17.2% of Facebook users update their status in an average day.

- For percent commenting on another's post, we have

$$0.22 \pm 1.96 \cdot \sqrt{\frac{0.22(0.78)}{970}} = 0.22 \pm 0.026 = \text{the interval from } 0.194 \text{ to } 0.246.$$

  We are 95% confident that between 19.4% and 24.6% of Facebook users comment on another's post in an average day.

- For percent commenting on another's photo, we have

$$0.20 \pm 1.96 \cdot \sqrt{\frac{0.20(0.80)}{970}} = 0.20 \pm 0.025 = \text{the interval from } 0.175 \text{ to } 0.225.$$

  We are 95% confident that between 17.5% and 22.5% of Facebook users comment on another's photo in an average day.

- For percent "liking" another's content, we have

$$0.26 \pm 1.96 \cdot \sqrt{\frac{0.26(0.74)}{970}} = 0.26 \pm 0.028 = \text{the interval from } 0.232 \text{ to } 0.288.$$

  We are 95% confident that between 23.2% and 28.8% of Facebook users "like" another's content in an average day.

- For percent sending another user a private message, we have

$$0.10 \pm 1.96 \cdot \sqrt{\frac{0.10(0.90)}{970}} = 0.10 \pm 0.019 = \text{the interval from } 0.081 \text{ to } 0.119.$$

  We are 95% confident that between 8.1% and 11.9% of Facebook users send another user a private message in an average day.

(b) The plausible proportions for those commenting on another's content are those between 0.194 and 0.246, while the plausible proportions for those updating their status are those between 0.128 to 0.172. Since these ranges do not overlap, we can be relatively confident that these proportions are not the same. A greater percentage comment on another's content than update their own status.

6.27 For 95% confidence, we have $z^* = 1.96$, so the margin of error for estimating the proportion of Democrats is

$$ME = z^* \cdot \sqrt{\frac{\hat{p}(1-\hat{p})}{n}} = 1.96 \cdot \sqrt{\frac{0.290(0.710)}{12000}} = 0.0081.$$

The margin of error for estimating the proportion of Republicans is

$$ME = z^* \cdot \sqrt{\frac{\hat{p}(1-\hat{p})}{n}} = 1.96 \cdot \sqrt{\frac{0.260(0.740)}{12000}} = 0.0078.$$

The proportion of Democrats might be as low as $0.290 - .0081 = 0.2819$ while the proportion of Republicans might be as high as $0.260 + 0.0078 = 0.2678$. Even at the extremes of the confidence intervals, the proportion of Democrats is still higher. Thus we can feel comfortable concluding that more American adults self-identified as Democrats than as Republicans in February 2010.

6.29 Using *StatKey* or other technology to create a bootstrap distribution, we see for one set of 1000 simulations that $SE = 0.045$. (Answers may vary slightly with other simulations.) Using the formula from the Central Limit Theorem, and using $\hat{p} = 0.583$ as an estimate for p, we have

$$SE = \sqrt{\frac{p(1-p)}{n}} \approx \sqrt{\frac{0.583(1-0.583)}{120}} = 0.045$$

We see that the bootstrap standard error and the formula match very closely.

6.31 Using *StatKey* or other technology to create a bootstrap distribution, we see for one set of 1000 simulations that $SE = 0.014$. (Answers may vary slightly with other simulations.) Using the formula from the Central Limit Theorem, and using $\hat{p} = 0.753$ as an estimate for p, we have

$$SE = \sqrt{\frac{p(1-p)}{n}} \approx \sqrt{\frac{0.753(1-0.753)}{1000}} = 0.014$$

We see that the bootstrap standard error and the formula match very closely.

6.33 Using *StatKey* or other technology we create a bootstrap distribution with at least 1000 simulated proportions. To find a 95% confidence interval we find the endpoints that contain 95% of the simulated proportions. For one set of 1000 bootstrap proportions we find that a 95% confidence interval for the proportion of orange Reese's Pieces goes from 0.40 to 0.56.

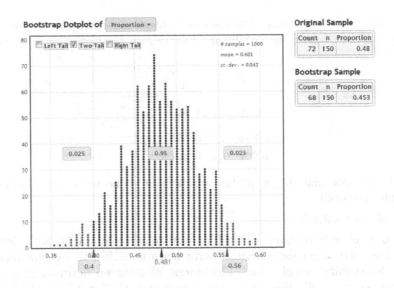

Using the normal distribution and the formula for standard error, we have

$$0.48 \pm 1.96 \cdot \sqrt{\frac{0.48(0.52)}{150}} = 0.48 \pm 0.080 = (0.40, 0.56).$$

In this case, the two methods give exactly the same interval.

6.35 We have $ME = 0.03$ for the margin of error. Since we are given no information about the population proportion, we use the conservative estimate $\tilde{p} = 0.5$.

For 99% confidence, we use $z^* = 2.576$. We have:

$$n = \left(\frac{z^*}{ME}\right)^2 \tilde{p}(1 - \tilde{p}) = \left(\frac{2.576}{0.03}\right)^2 (0.5 \cdot 0.5) = 1843.3$$

We round up to $n = 1844$.

For 95% confidence, we use $z^* = 1.96$. We have:

$$n = \left(\frac{z^*}{ME}\right)^2 \tilde{p}(1 - \tilde{p}) = \left(\frac{1.96}{0.03}\right)^2 (0.5 \cdot 0.5) = 1067.1$$

We round up to $n = 1068$.

For 90% confidence, we use $z^* = 1.645$. We have:

$$n = \left(\frac{z^*}{ME}\right)^2 \tilde{p}(1 - \tilde{p}) = \left(\frac{1.645}{0.03}\right)^2 (0.5 \cdot 0.5) = 751.7$$

We round up to $n = 752$.

We see that the sample size goes up as the level of confidence we want in the result goes up. Or, put another way, a larger sample size gives a higher level of confidence in the accuracy of the estimate.

6.37 (a) The sample size is definitely large enough to use the normal distribution. The relevant sample statistic is $\hat{p} = 0.32$ and, for a 95% confidence interval, we use $z^* = 1.96$. The confidence interval is

$$\hat{p} \quad \pm \quad z^* \sqrt{\frac{\hat{p}(1 - \hat{p})}{n}}$$

$$0.32 \quad \pm \quad 1.96 \cdot \sqrt{\frac{0.32(1 - 0.32)}{1000}}$$

$$0.32 \quad \pm \quad 0.029$$

$$0.291 \quad \text{to} \quad 0.349$$

We are 95% confident that the proportion of US adults who favor a tax on soda and junk food is between 0.291 and 0.349.

(b) The margin of error is 2.9%.

(c) Since the margin of error is about ±3% with a sample size of $n = 1000$, we'll definitely need a sample size larger than 1000 to get the margin of error down to ±1%. To see how much larger, we use the formula for determining sample size. The margin of error we desire is $ME = 0.01$, and for 95% confidence we use $z^* = 1.96$. We can use the sample statistic $\hat{p} = 0.32$ as our best estimate for p. We have:

$$n \quad = \quad \left(\frac{z^*}{ME}\right)^2 \tilde{p}(1 - \tilde{p})$$

$$= \quad \left(\frac{1.96}{0.01}\right)^2 0.32(1 - 0.32)$$

$$= \quad 8359.3.$$

We round up, so we would need to include 8,360 people in the survey in order to get the margin of error down to within ±1%.

6.39 Since we have no reasonable estimate for the proportion, we use $\tilde{p} = 0.5$. For 98% confidence, use $z^* = 2.326$. The required sample size is

$$n = \left(\frac{2.326}{0.04}\right)^2 0.5(1 - 0.5) = 845.4$$

We round up to show a sample of at least 846 individuals is needed to estimate the proportion who would consider buying a sunscreen pill to within 4% with 98% confidence.

6.41 We have $ME = 0.02$ so the sample size needed is $n = 1/(0.02)^2 = 2500$.

6.43 We have $ME = 0.05$ so the sample size needed is $n = 1/(0.05)^2 = 400$.

6.45 We see in the dataset **ICUAdmissions** that 160 of the patients in the sample lived and 40 died. The sample proportion who live is $\hat{p} = 160/200 = 0.80$. Since there are more than 10 in the living and the dying groups in the sample, we may use the normal approximation to construct a confidence interval for the proportion who live. For 95% confidence the standard normal endpoint is $z^* = 1.96$, so we compute the interval with

$$0.80 \pm 1.96\sqrt{\frac{0.80(1 - 0.80)}{200}} = 0.80 \pm 0.055 = (0.745, 0.855).$$

We are 95% sure that the proportion of ICU patients (at this hospital) who live is between 74.5% and 85.5%.

Section 6.1-HT Solutions

6.47 Since $np_0 = 200(0.3) = 60$ and $n(1 - p_0) = 200(0.7) = 140$, the sample size is large enough to use the normal distribution. In general, the standardized test statistic is

$$z = \frac{\text{Sample Statistic} - \text{Null Parameter}}{SE}.$$

In this test for a proportion, the sample statistic is $\hat{p} = 0.21$ and the parameter from the null hypothesis is $p_0 = 0.3$. The standard error is $SE = \sqrt{p_0(1 - p_0)/n}$. The standardized test statistic is

$$z = \frac{\hat{p} - p_0}{\sqrt{\frac{p_0(1-p_0)}{n}}} = \frac{0.21 - 0.3}{\sqrt{\frac{0.3(0.7)}{200}}} = -2.78.$$

This is a lower-tail test, so the p-value is the area below -2.78 in a standard normal distribution. Using technology or a table, we see that the p-value is 0.0027. This p-value is very small so we find strong evidence to support the alternative hypothesis that $p < 0.3$.

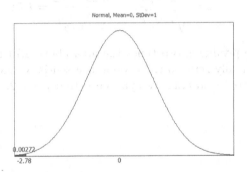

6.49 Since $np_0 = 50(0.8) = 40$ and $n(1 - p_0) = 50(0.2) = 10$, the sample size is (just barely) large enough to use the normal distribution. In general, the standardized test statistic is

$$z = \frac{\text{Sample Statistic} - \text{Null Parameter}}{SE}.$$

In this test for a proportion, the sample statistic is $\hat{p} = 0.88$ and the parameter from the null hypothesis is $p_0 = 0.8$. The standard error is $SE = \sqrt{p_0(1 - p_0)/n}$. The standardized test statistic is

$$z = \frac{\hat{p} - p_0}{\sqrt{\frac{p_0(1-p_0)}{n}}} = \frac{0.88 - 0.80}{\sqrt{\frac{0.8(0.2)}{50}}} = 1.41.$$

This is an upper-tail test, so the p-value is the area above 1.41 in a standard normal distribution. Using technology or a table, we see that the p-value is 0.079. This p-value is larger than the significance level of 5%, so we do not find sufficient evidence to support the alternative hypothesis that $p > 0.8$.

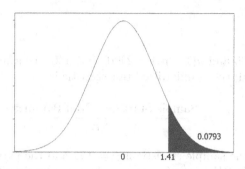

6.51 Since $np_0 = 1000(0.2) = 200$ and $n(1 - p_0) = 1000(0.8) = 800$, the sample size is large enough to use the normal distribution. In general, the standardized test statistic is

$$z = \frac{\text{Sample Statistic} - \text{Null Parameter}}{SE}.$$

In this test for a proportion, the sample statistic is $\hat{p} = 0.26$ and the parameter from the null hypothesis is $p_0 = 0.2$. The standard error is $SE = \sqrt{p_0(1 - p_0)/n}$. The standardized test statistic is

$$z = \frac{\hat{p} - p_0}{\sqrt{\frac{p_0(1 - p_0)}{n}}} = \frac{0.26 - 0.2}{\sqrt{\frac{0.2(0.8)}{1000}}} = 4.74.$$

This is a two-tail test, so the p-value is two times the area above 4.74 in a standard normal distribution. The area beyond 4.74 is essentially zero, so the p-value is essentially zero. The p-value is very small so we have strong evidence to support the alternative hypothesis that $p \neq 0.2$.

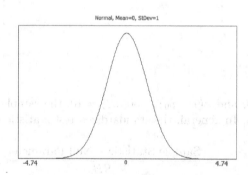

6.53 (a) If we let p_l be the proportion of left-handed lawyers, then we are testing test $H_0 : p_l = 0.10$ vs. $H_a : p_l \neq 0.10$.

   (b) The sample proportion is $\hat{p}_l = 16/105 = 0.1524$. The test statistic is

$$z = \frac{0.1524 - 0.10}{\sqrt{\frac{0.10(1 - .10)}{105}}} = 1.79.$$

The area above 1.79 in the standard normal curve is 0.037, so the p-value is $2(0.037) = 0.074$.

(c) We do not reject H_0 at the 5% significance level, and thus do not conclude that the proportion of left-handed lawyers differs from the proportion of left-handed Americans. At the 10% significance level we do reject H_0 and conclude that there is a higher percentage of left-handed lawyers.

6.55 We are conducting a hypothesis test for a proportion p, where p is the proportion of all MLB games won by the home team. We are testing to see if there is evidence that $p > 0.5$, so we have

$$H_0 : \quad p = 0.5$$
$$H_a : \quad p > 0.5$$

This is a one-tail test since we are specifically testing to see if the proportion is greater than 0.5. The test statistic is:

$$z = \frac{\text{Sample statistic} - \text{Null parameter}}{SE} = \frac{\hat{p} - p_0}{\sqrt{\frac{p_0(1-p_0)}{n}}} = \frac{0.549 - 0.5}{\sqrt{\frac{0.5(0.5)}{2430}}} = 4.83.$$

Using the normal distribution, we find a p-value of (to five decimal places) zero. This provides very strong evidence to reject H_0 and conclude that the home team wins more than half the games played. The home field advantage is real!

6.57 We are conducting a hypothesis test for a proportion p, where p is the proportion of all US adults who know most or all of their neighbors. We are testing to see if there is evidence that $p > 0.5$, so we have

$$H_0 : \quad p = 0.5$$
$$H_a : \quad p > 0.5$$

This is a one-tail test since we are specifically testing to see if the proportion is greater than 0.5. The sample proportion is $\hat{p} = 0.51$ and the null proportion is $p_0 = 0.5$. The sample size is $n = 2,255$ so the test statistic is:

$$z = \frac{\text{Sample statistic} - \text{Null parameter}}{SE} = \frac{\hat{p} - p_0}{\sqrt{\frac{p_0(1-p_0)}{n}}} = \frac{0.51 - 0.5}{\sqrt{\frac{0.5(0.5)}{2255}}} = 0.95.$$

This is a one-tail test, so the p-value is the area above 0.95 in the standard normal distribution. We find a p-value of 0.171. This p-value is larger than even a 10% significance level, so we do not have sufficient evidence to conclude that the proportion of US adults who know their neighbors is larger than 0.5.

6.59 Let p be the proportion of times Team B will beat Team A. We are testing $H_0 : p = 0.5$ vs $H_a : p \neq 0.5$. The sample proportion is $\hat{p} = 24/40 = 0.6$. The test statistic is

$$z = \frac{0.6 - 0.5}{\sqrt{\frac{0.5(1-0.5)}{40}}} = 1.26$$

The area to the right of 1.26 on the standard normal distribution is 0.104, so the p-value is $2(0.104) = 0.208$. There is not convincing evidence that one team is better than the other. (We arrive at the same conclusion if we let p be the proportion of times that Team A wins.)

6.61 We use technology to determine that the number of smokers in the sample is 43, so the sample proportion of smokers is $\hat{p} = 43/315 = 0.1365$. The hypotheses are:

$$H_0 : \quad p = 0.20$$
$$H_a : \quad p \neq 0.20$$

The test statistic is:

$$z = \frac{\text{Sample statistic} - \text{Null parameter}}{SE} = \frac{\hat{p} - p_0}{\sqrt{\frac{p_0(1-p_0)}{n}}} = \frac{0.1365 - 0.20}{\sqrt{\frac{0.2(0.8)}{315}}} = -2.82.$$

This is a two-tail test, so the p-value is twice the area below -2.82 in a standard normal distribution. We see that the p-value is $2(0.0024) = 0.0048$. This small p-value leads us to reject H_0. We find strong evidence that the proportion of smokers is not 20%.

Section 6.2-D Solutions

6.63 The sample means will have a standard error of

$$SE = \frac{\sigma}{\sqrt{n}} = \frac{5}{\sqrt{1000}} = 0.158.$$

6.65 The sample means will have a standard error of

$$SE = \frac{\sigma}{\sqrt{n}} = \frac{80}{\sqrt{40}} = 12.65.$$

6.67 We use a t-distribution with df = 9. Using technology, we see that the values with 5% beyond them in each tail are ±1.83.

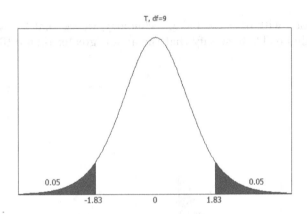

6.69 We use a t-distribution with df = 24. Using technology, we see that the values with 0.025 beyond them in each tail are ±2.06.

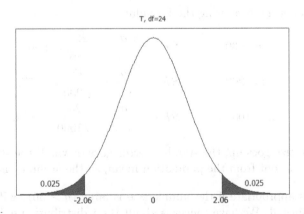

6.71 We use a t-distribution with df = 5. Using technology, we see that the area above 2.3 is 0.0349. (On a paper table, we may only be able to specify that the area is between 0.025 and 0.05.)

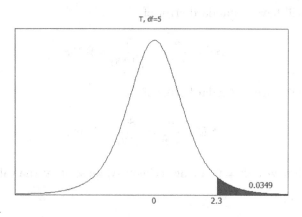

6.73 We use a t-distribution with df = 19. Using technology, we see that the area below -1.0 is 0.165. (On a paper table, we may only be able to specify that the area is greater than 0.10.)

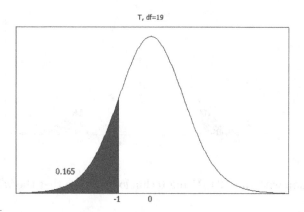

6.75 We compute the standard errors using the formula:

$$n = 30: \qquad SE = \frac{\sigma}{\sqrt{n}} = \frac{25}{\sqrt{30}} = 4.56.$$

$$n = 200: \qquad SE = \frac{\sigma}{\sqrt{n}} = \frac{25}{\sqrt{200}} = 1.77.$$

$$n = 1000: \qquad SE = \frac{\sigma}{\sqrt{n}} = \frac{25}{\sqrt{1000}} = 0.79.$$

We see that as the sample size goes up, the standard error goes down. If the standard error goes down, the sample means are less spread out from the population mean, so the accuracy is better.

6.77 The t-distribution is appropriate if the sample size is large ($n \geq 30$) or if the underlying distribution appears to be relatively normal. We have concerns about the t-distribution only for small sample sizes and heavy skewness or outliers. In this case, the sample size is small ($n = 12$) but the distribution is not heavily skewed and it does not have extreme outliers. A condition of normality is reasonable, so the t-distribution is appropriate. For the degrees of freedom df and estimated standard error SE, we have:

$$df = n - 1 = 12 - 1 = 11, \quad \text{and} \quad SE = \frac{s}{\sqrt{n}} = \frac{1.6}{\sqrt{12}} = 0.46.$$

6.79 The t-distribution is appropriate if the sample size is large ($n \geq 30$) or if the underlying distribution appears to be relatively normal. We have concerns about the t-distribution only for small sample sizes and heavy skewness or outliers. In this case, the sample size is small ($n = 18$) and the data is heavily skewed with some apparent outliers. It would not be appropriate to use the t-distribution in this case. We might try analyzing the data using simulation methods such as a bootstrap or randomization distribution.

Section 6.2-CI Solutions

6.81 For a confidence interval for μ using the t-distribution, we use

$$\bar{x} \pm t^* \cdot \frac{s}{\sqrt{n}}$$

We use a t-distribution with df = 29, so for a 95% confidence interval, we have $t^* = 2.05$. The confidence interval is

$$12.7 \quad \pm \quad 2.05 \cdot \frac{5.6}{\sqrt{30}}$$
$$12.7 \quad \pm \quad 2.10$$
$$10.6 \quad \text{to} \quad 14.8$$

The best estimate for μ is $\bar{x} = 12.7$, the margin of error is ± 2.10, and the 95% confidence interval for μ is 10.6 to 14.8. We are 95% confident that the mean of the entire population is between 10.6 and 14.8.

6.83 For a confidence interval for μ using the t-distribution, we use

$$\bar{x} \pm t^* \cdot \frac{s}{\sqrt{n}}$$

We use a t-distribution with df = 99, so for a 90% confidence interval, we have $t^* = 1.66$. (Since the sample size is so large, the t distribution value is almost identical to the standard normal z value.) The confidence interval is

$$3.1 \quad \pm \quad 1.66 \cdot \frac{0.4}{\sqrt{100}}$$
$$3.1 \quad \pm \quad 0.066$$
$$3.034 \quad \text{to} \quad 3.166$$

The best estimate for μ is $\bar{x} = 3.1$, the margin of error is ± 0.066, and the 90% confidence interval for μ is 3.034 to 3.166. We are 90% confident that the mean of the entire population is between 3.034 and 3.166.

6.85 For a confidence interval for μ using the t-distribution, we use

$$\bar{x} \pm t^* \cdot \frac{s}{\sqrt{n}}$$

We use a t-distribution with df = 9, so for a 99% confidence interval, we have $t^* = 3.25$. The confidence interval is

$$46.1 \quad \pm \quad 3.25 \cdot \frac{12.5}{\sqrt{10}}$$
$$46.1 \quad \pm \quad 12.85$$
$$33.25 \quad \text{to} \quad 58.95$$

The best estimate for μ is $\bar{x} = 46.1$, the margin of error is ± 12.85, and the 99% confidence interval for μ is 33.25 to 58.95. We are 99% confident that the mean of the entire population is between 33.35 and 58.95.

6.87 The desired margin of error is $ME = 5$ and we have $z^* = 1.96$ for 95% confidence. We use $\tilde{\sigma} = 18$ to approximate the standard deviation. We use the formula to find sample size:

$$n = \left(\frac{z^* \cdot \tilde{\sigma}}{ME}\right)^2 = \left(\frac{1.96 \cdot 18}{5}\right)^2 = 49.8$$

We round up to $n = 50$. In order to ensure that the margin of error is within the desired ±5 units, we should use a sample size of 50 or higher.

6.89 The desired margin of error is $ME = 0.5$ and we have $z^* = 1.645$ for 90% confidence. We use $\tilde{\sigma} = 25$ to approximate the standard deviation. We use the formula to find sample size:

$$n = \left(\frac{z^* \cdot \tilde{\sigma}}{ME}\right)^2 = \left(\frac{1.645 \cdot 25}{0.5}\right)^2 = 6765.1$$

We round up to $n = 6766$. In order to ensure that the margin of error is within the desired ±0.5 units, we would need to use a sample size of 6,766 or higher.

6.91 We have $n = 55$ and $\bar{x} = 2.4$ and $s = 1.51$. Using $df = 54$ and a 99% confidence level, we see that $t^* = 2.670$. The 99% confidence interval is given by:

$$
\begin{aligned}
Statistic \quad &\pm \quad z^* \cdot SE \\
\bar{x} \quad &\pm \quad t^* \cdot \frac{s}{\sqrt{n}} \\
2.4 \quad &\pm \quad 2.670 \cdot \frac{1.51}{\sqrt{55}} \\
2.4 \quad &\pm \quad 0.544 \\
1.856 \quad &\text{to} \quad 2.944.
\end{aligned}
$$

We are 99% sure that the mean number of kills per week for all US household cats is between 1.856 and 2.944.

6.93 (a) The margin of error for estimating μ is given by

$$ME = t^* \cdot \frac{s}{\sqrt{n}}$$

For a t-distribution with df = 2,005 and 99% confidence, we have $t^* = 2.578$. The margin of error is

$$ME = t^* \cdot \frac{s}{\sqrt{n}} = 2.578 \cdot \frac{1.4}{\sqrt{2006}} = 0.08$$

With 99% confidence, the best estimate of the average number of close confidants is 2.2 with a margin of error of 0.08.

(b) The 99% confidence interval is the best estimate plus/minus the margin of error. We see that the 99% confidence interval is

$$\overline{x} \pm ME = 2.2 \pm 0.08 = 2.12 \text{ to } 2.28$$

We are 99% sure that the average number of close confidants for US adults is between 2.12 and 2.28.

6.95 We use a t-distribution with df = 98, so for a 95% confidence interval, we have $t^* = 1.98$. The confidence interval is

$$
\begin{array}{rcl}
\overline{x} & \pm & t^* \cdot \dfrac{s}{\sqrt{n}} \\[2mm]
564 & \pm & 1.98 \cdot \dfrac{122}{\sqrt{99}} \\[2mm]
564 & \pm & 24.3 \\[2mm]
539.7 & \text{to} & 588.3
\end{array}
$$

We are 95% sure that the mean number of unique genes in the gut bacteria of European individuals is between 539.7 and 588.3 million.

6.97 The sample size of $n = 9$ is quite small, so we require a condition of approximate normality for the underlying population in order to use the t-distribution. In the dotplot of the data, it appears that the data might be right skewed and there is quite a large outlier. It is probably more reasonable to use other methods, such as a bootstrap distribution, to compute a confidence interval using this data.

6.99 Using *StatKey* or other technology to create a bootstrap distribution, we see for one set of 1000 simulations that $SE \approx 5.46$. (Answers may vary slightly with other simulations.) Using the formula from the Central Limit Theorem, and using $s = 27.26$ as an estimate for σ, we have

$$SE = \frac{s}{\sqrt{n}} = \frac{27.26}{\sqrt{24}} = 5.56.$$

We see that the bootstrap standard error and the formula match relatively closely.

6.101 Using *StatKey* or other technology to create a bootstrap distribution, we see for one set of 1000 simulations that $SE \approx 2.1$. (Answers may vary slightly with other simulations.) Using the formula from the Central Limit Theorem, and using $s = 11.11$ as an estimate for σ, we have

$$SE = \frac{s}{\sqrt{n}} = \frac{11.11}{\sqrt{25}} = 2.22.$$

We see that the bootstrap standard error and the formula match very closely.

6.103 We use *StatKey* or other technology to create a bootstrap distribution with at least 1000 simulated means from samples of the Atlanta commute distances. To find a 95% confidence interval we find the endpoints that contain 95% of the simulated means. For one set of 1000 bootstrap means shown below we find that a 95% confidence interval for mean Atlanta commute distance goes from 16.92 to 19.34 miles.

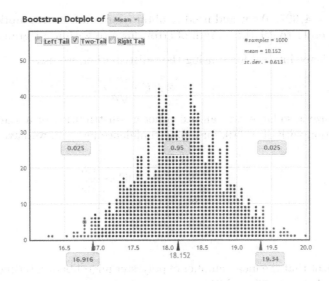

For a 95% confidence interval with $df = 499$, we have $t^* = 1.965$. Using the t-distribution and the formula for standard error, we have

$$18.156 \pm 1.965 \cdot \frac{13.798}{\sqrt{500}} = 18.156 \pm 1.21 = (16.95, 19.37)$$

The two methods give very similar intervals.

6.105 (a) For a confidence interval for μ using the t-distribution, we use

$$\bar{x} \pm t^* \cdot \frac{s}{\sqrt{n}}$$

Using a t-distribution with df = 314, for a 95% confidence interval, we have $t^* = 1.97$. The confidence interval is

$$
\begin{array}{rcl}
77.03 & \pm & 1.97 \cdot \dfrac{33.83}{\sqrt{315}} \\
77.03 & \pm & 3.76 \\
73.27 & \text{to} & 80.79
\end{array}
$$

We are 95% confident that the average number of grams of fat consumed per day by all US adults is between 73.27 grams and 80.79 grams.

(b) The margin of error is ± 3.76 grams of fat.

(c) Since the margin of error is ± 3.76 with a sample size of $n = 315$, we'll definitely need a sample size larger than 315 to get the margin of error down to ± 1. To see how much larger, we use the formula for determining sample size. The margin of error we desire is $ME = 1$, and for 95% confidence we use $z^* = 1.96$. We can use the sample statistic $s = 33.83$ as our best estimate for σ. We have:

$$n = \left(\frac{z^* \cdot \tilde{\sigma}}{ME} \right)^2 = \left(\frac{1.96 \cdot 33.83}{1} \right)^2 = 4396.6$$

We round up to $n = 4,397$. We would need to obtain data on fat consumption from a sample of $4,397$ or more people in order to get the margin of error down to within ± 1 gram.

6.107 (a) For a confidence interval for μ using the t-distribution, we use

$$\bar{x} \pm t^* \cdot \frac{s}{\sqrt{n}}$$

Since the sample size is so large, we can use either a t-distribution or a standard normal distribution. Using a t-distribution with df $= 119$, for a 99% confidence interval, we have $t^* = 2.62$. The confidence interval is

$$
\begin{array}{rcl}
290 & \pm & 2.62 \cdot \dfrac{87.6}{\sqrt{120}} \\
290 & \pm & 21.0 \\
269.0 & \text{to} & 311.0
\end{array}
$$

We are 99% confident that the mean number of polyester microfibers entering wastewater when washing a fleece garment is between 269 and 311 per liter.

(b) The margin of error is ± 21.0 microfibers per liter.

(c) Since the margin of error is ± 21.0 with a sample size of $n = 120$, we'll definitely need a sample size larger than 120 to get the margin of error down to ± 5. To see how much larger, we use the formula for determining sample size. The margin of error we desire is $ME = 5$, and for 99% confidence we use $z^* = 2.576$. We can use the sample statistic $s = 87.6$ as our best estimate for σ. We have:

$$n = \left(\frac{z^* \cdot \tilde{\sigma}}{ME}\right)^2 = \left(\frac{2.576 \cdot 87.6}{5}\right)^2 = 2036.85$$

We round up to $n = 2037$. We would need to obtain at least 2037 samples of wastewater after washing fleece to get the margin of error down to ± 5 particles per liter of wastewater.

6.109 We use $z^* = 1.96$ for 95% confidence, and we use $\tilde{\sigma} = 30$.

For a desired margin of error of $ME = 10$, we have:

$$n = \left(\frac{z^* \cdot \tilde{\sigma}}{ME}\right)^2 = \left(\frac{1.96 \cdot 30}{10}\right)^2 = 34.6$$

We round up to $n = 35$.

For a desired margin of error of $ME = 5$, we have:

$$n = \left(\frac{z^* \cdot \tilde{\sigma}}{ME}\right)^2 = \left(\frac{1.96 \cdot 30}{5}\right)^2 = 138.3$$

We round up to $n = 139$.

For a desired margin of error of $ME = 1$, we have:

$$n = \left(\frac{z^* \cdot \tilde{\sigma}}{ME}\right)^2 = \left(\frac{1.96 \cdot 30}{1}\right)^2 = 3457.4$$

We round up to $n = 3,458$.

We see that the sample size goes up as we require more accuracy. Or, put another way, a larger sample size gives greater accuracy.

6.111 We use $z^* = 1.96$ for 95% confidence, and we want a margin of error of $ME = 3$.

Using $\tilde{\sigma} = 100$, we have:
$$n = \left(\frac{z^* \cdot \tilde{\sigma}}{ME}\right)^2 = \left(\frac{1.96 \cdot 100}{3}\right)^2 = 4268.4$$

We round up to $n = 4{,}269$.

Using $\tilde{\sigma} = 50$, we have:
$$n = \left(\frac{z^* \cdot \tilde{\sigma}}{ME}\right)^2 = \left(\frac{1.96 \cdot 50}{3}\right)^2 = 1067.1$$

We round up to $n = 1{,}068$.

Using $\tilde{\sigma} = 10$, we have:
$$n = \left(\frac{z^* \cdot \tilde{\sigma}}{ME}\right)^2 = \left(\frac{1.96 \cdot 10}{3}\right)^2 = 42.7$$

We round up to $n = 43$.

Not surprisingly, we see that the more variability there is in the underlying data, the larger sample size we need to get the accuracy we want. If the original data are very spread out, we need a large sample size to get an accurate estimate. If, however, the original data are very narrowly focused, a smaller sample size will do the trick.

6.113 Using any statistics package, we see that a 95% confidence interval is 12.79 ± 0.59 12.20 to 13.38. The average number of grams of fiber eaten in a day is between 12.20 grams and 13.38 grams.

Section 6.2-HT Solutions

6.115 In general, the standardized test statistic is

$$\frac{\text{Sample Statistic} - \text{Null Parameter}}{SE}.$$

In this test for a mean, the sample statistic is $\bar{x} = 91.7$ and the parameter from the null hypothesis is $\mu_0 = 100$. The standard error is $SE = s/\sqrt{n}$. The standardized test statistic is

$$t = \frac{\bar{x} - \mu_0}{s/\sqrt{n}} = \frac{91.7 - 100}{12.5/\sqrt{30}} = -3.64.$$

This is a lower-tail test, so the p-value is the area below -3.64 in a t-distribution with df $= 29$. We see that the p-value is 0.0005. This p-value is very small, and below any reasonable significance level. There is strong evidence to support the alternative hypothesis that $\mu < 100$.

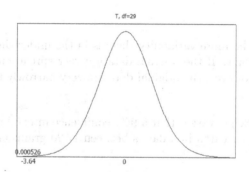

6.117 This sample size is quite small, but we are told that the underlying distribution is approximately normal so we can proceed with the t-test. In general, the standardized test statistic is

$$\frac{\text{Sample Statistic} - \text{Null Parameter}}{SE}.$$

In this test for a mean, the sample statistic is $\bar{x} = 13.2$ and the parameter from the null hypothesis is $\mu_0 = 10$. The standard error is $SE = s/\sqrt{n}$. The standardized test statistic is

$$t = \frac{\bar{x} - \mu_0}{s/\sqrt{n}} = \frac{13.2 - 10}{8.7/\sqrt{12}} = 1.27.$$

This is an upper-tail test, so the p-value is the area above 1.27 in a t-distribution with df $= 11$. We see that the p-value is 0.115. This p-value is larger than the significance level of 0.05 (and larger than any reasonable significance level), so we do not reject H_0 and do not find sufficient evidence to support the alternative hypothesis that $\mu > 10$.

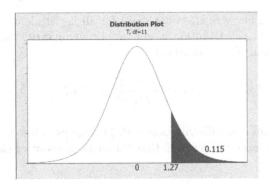

6.119 In general, the standardized test statistic is

$$\frac{\text{Sample Statistic} - \text{Null Parameter}}{SE}.$$

In this test for a mean, the sample statistic is $\bar{x} = 432$ and the parameter from the null hypothesis is $\mu_0 = 500$. The standard error is $SE = s/\sqrt{n}$. The standardized test statistic is

$$t = \frac{\bar{x} - \mu_0}{s/\sqrt{n}} = \frac{432 - 500}{118/\sqrt{75}} = -4.99.$$

This is a two-tail test, so the p-value is two times the area below -4.99 in a t-distribution with df = 74. We see that the p-value is essentially zero. This p-value is extremely small and provides strong evidence to support the alternative hypothesis that $\mu \neq 500$.

6.121 The sample size of $n = 7$ is very small so it is important in using the t-distribution to know that the values are not heavily skewed. The mean for non-autistic male children is 1.15 billion, so the null and alternative hypotheses are

$$H_0: \quad \mu = 1.15$$
$$H_a: \quad \mu > 1.15$$

where μ represents the mean number of neurons, in billions, in the prefrontal cortex for male autistic children. In general, the standardized test statistic is

$$\frac{\text{Sample Statistic} - \text{Null Parameter}}{SE}.$$

In this test for a mean, the sample statistic is $\bar{x} = 1.94$ and the parameter from the null hypothesis is $\mu_0 = 1.15$. The standard error is $SE = s/\sqrt{n}$. The standardized test statistic is

$$t = \frac{\bar{x} - \mu_0}{s/\sqrt{n}} = \frac{1.94 - 1.15}{0.50/\sqrt{7}} = 4.18.$$

This is an upper-tail test, so the p-value is the area above 4.18 in a t-distribution with df = 6. Using technology we see that the p-value is 0.003. This p-value is very small so we reject the null hypothesis. There is strong evidence that, on average, male autistic children have an overabundance of neurons in the prefrontal cortex.

6.123 Let μ denote the mean hours of sleep per night for all her students. We wish to test the hypotheses $H_0 : \mu = 8$ vs $H_a : \mu < 8$. The distribution is not exactly symmetric, but at least has no outliers, so we proceed with the t-distribution. The t-statistic is

$$t = \frac{6.2 - 8}{1.7/\sqrt{12}} = -3.67$$

The p-value from the lower tail of a t-distribution with 11 degrees of freedom is 0.0018. This is a very small p-value, providing strong evidence to conclude that the mean amount of sleep for her students is less than 8 hours.

6.125 In each of parts (a), (b), and (c), the hypotheses are:

$$H_0 : \quad \mu = 265$$
$$H_a : \quad \mu > 265$$

where μ stands for the mean cost of a house in the relevant state, in thousands of dollars.

(a) For New York, we calculate the test statistic

$$t = \frac{\bar{x} - \mu_0}{s/\sqrt{n}} = \frac{565.6 - 265}{697.6/\sqrt{30}} = 2.36$$

Using an upper-tail test with a t-distribution with 29 degrees of freedom, we get a p-value of 0.0126. At a 5% level, we reject the null hypothesis. The average cost of a house in New York State is significantly more then the US average.

(b) For New Jersey, we calculate the test statistic

$$t = \frac{\bar{x} - \mu_0}{s/\sqrt{n}} = \frac{388.5 - 265}{224.7/\sqrt{30}} = 3.01$$

Using an upper-tail test with a t-distribution with 29 degrees of freedom, we get a p-value of 0.0027. We reject the null hypothesis and find strong evidence that the average cost of a house in New Jersey is significantly more then the US average.

(c) Note that the sample mean for Pennsylvania, $\bar{x} = 249.6$ is less than the US average, $\mu = 265$. We don't need a statistical test to see that the average cost in Pennsylvania is not significantly greater than the US average. However if we conducted the test we would have

$$t = \frac{\bar{x} - \mu_0}{s/\sqrt{n}} = \frac{249.6 - 265}{179.3/\sqrt{30}} = -0.47$$

Using an upper-tail test with a t-distribution with 29 degrees of freedom, we get a p-value of 0.6791. (Notice that we still use the upper-tail, which gives a p-value in this case that is greater than 0.5.)

(d) New Jersey has the most evidence that the state average is greater then the US average (smallest p-value). This may surprise you, since the mean for NY homes is greater. However, the standard deviation is also important in determining evidence for or against a claim.

6.127   (a) We see that $n = 30$ with $\bar{x} = 0.2511$ and $s = 0.01096$.

(b) The hypotheses are:

$$H_0: \quad \mu = 0.260$$
$$H_a: \quad \mu \neq 0.260$$

where μ represents the mean of all team batting averages in Major League Baseball. We calculate the test statistic

$$t = \frac{\bar{x} - \mu_0}{s/\sqrt{n}} = \frac{0.2511 - 0.260}{0.01096/\sqrt{30}} = -4.45$$

We use a t-distribution with 29 degrees of freedom to see that the proportion below -4.45 is 0.000058. Since this is a two-tail test, the p-value is $2(0.000058) = 0.000116$. We reject the null hypothesis, and conclude that the average team batting average is different from (and less than) 0.260.

(c) The test statistic matches the computer output exactly and the p-value is the same up to rounding off.

6.129   (a) The hypotheses are:

$$H_0: \quad \mu = 1.0$$
$$H_a: \quad \mu < 1.0$$

where μ represents the mean mercury level of fish in all Florida lakes. Some computer output for the test is shown:

```
One-Sample T: Avg_Mercury
Test of mu = 1 vs < 1
                                              95% Upper
Variable      N    Mean    StDev   SE Mean     Bound       T       P
Avg_Mercury  53   0.5272  0.3410   0.0468      0.6056   -10.09   0.000
```

We see that the p-value is approximately 0, so there is strong evidence that the mean mercury level of fish in Florida lakes is less than 1.0 ppm.

(b) The hypotheses are:

$$H_0: \quad \mu = 0.5$$
$$H_a: \quad \mu < 0.5$$

where μ represents the mean mercury level of fish in all Florida lakes. Some computer output for the test is shown:

```
One-Sample T: Avg_Mercury
Test of mu = 0.5 vs < 0.5
                                              95% Upper
Variable      N    Mean    StDev   SE Mean     Bound      T       P
Avg_Mercury  53   0.5272  0.3410   0.0468      0.6056   0.58   0.718
```

We see that the p-value is 0.718, so there is no evidence at all that the mean is less than 0.5. (In fact, we see that the sample mean, $\bar{x} = 0.5272$ ppm, is actually more than 0.5.)

Section 6.3-D Solutions

6.131 (a) The differences in sample proportions will have a standard error of

$$SE = \sqrt{\frac{p_A(1-p_A)}{n_A} + \frac{p_B(1-p_B)}{n_B}} = \sqrt{\frac{0.70(0.30)}{50} + \frac{0.60(0.40)}{75}} = 0.086$$

(b) We check the sample size for Group A: $n_A p_A = 50(0.70) = 35$ and $n_A(1-p_A) = 50(0.30) = 15$, and for Group B: $n_B p_B = 75(0.60) = 45$ and $n_B(1-p_B) = 75(0.40) = 30$. In both cases, the sample size is large enough and the normal distribution applies.

6.133 (a) The differences in sample proportions will have a standard error of

$$SE = \sqrt{\frac{p_A(1-p_A)}{n_A} + \frac{p_B(1-p_B)}{n_B}} = \sqrt{\frac{0.20(0.80)}{100} + \frac{0.30(0.70)}{50}} = 0.076$$

(b) We check the sample size for Group A: $n_A p_A = 100(0.20) = 20$ and $n_A(1-p_A) = 100(0.80) = 80$, and for Group B: $n_B p_B = 50(0.30) = 15$ and $n_B(1-p_B) = 50(0.7) = 35$. The sample sizes are large enough for the normal distribution to apply.

6.135 (a) The differences in sample proportions will have a standard error of

$$SE = \sqrt{\frac{p_A(1-p_A)}{n_A} + \frac{p_B(1-p_B)}{n_B}} = \sqrt{\frac{0.30(0.70)}{40} + \frac{0.24(0.76)}{30}} = 0.106$$

(b) We check the sample size for Group A: $n_A p_A = 40(0.30) = 12$ and $n_A(1-p_A) = 40(0.70) = 28$, and for Group B: $n_B p_B = 30(0.24) = 7.2$. Since $n_B p_B < 10$, the normal distribution does not apply in this case. For inference on the difference in sample proportions in this case, we should use bootstrap or randomization methods.

6.137 (a) This compares two proportions (PC user vs. Mac user) drawn from the same group (students). The methods of this section do not apply to this type of difference in proportions.

(b) This compares proportions (study abroad) for two different groups (public vs. private). The methods of this section are appropriate for this type of difference in proportions.

(c) This compares two proportions (in-state vs. out-of-state) drawn from the same group (students). The methods of this section do not apply to this type of difference in proportions.

(d) This compares proportions (get financial aid) from two different groups (in-state vs. out-of-state). The methods of this section are appropriate for this type of difference in proportions.

Section 6.3-CI Solutions

6.139 The sample sizes are both large enough to use the normal distribution. For a confidence interval using the normal distribution, we use

$$\text{Sample statistic} \pm z^* \cdot SE.$$

The relevant sample statistic for a confidence interval for a difference in proportions is $\hat{p}_1 - \hat{p}_2 = 0.72 - 0.68$. For a 95% confidence interval, we have $z^* = 1.96$, and we use the sample proportions in computing the standard error. The confidence interval is

$$
\begin{aligned}
(\hat{p}_1 - \hat{p}_2) \quad &\pm \quad z * \cdot \sqrt{\frac{\hat{p}_1(1-\hat{p}_1)}{n_1} + \frac{\hat{p}_2(1-\hat{p}_2)}{n_2}} \\
(0.72 - 0.68) \quad &\pm \quad 1.96 \cdot \sqrt{\frac{0.72(0.28)}{500} + \frac{0.68(0.32)}{300}} \\
0.04 \quad &\pm \quad 0.066 \\
-0.026 \quad &\text{to} \quad 0.106
\end{aligned}
$$

The best estimate for the difference in the two proportions $p_1 - p_2$ is 0.04, the margin of error is ± 0.066, and the 95% confidence interval for $p_1 - p_2$ is -0.026 to 0.106.

6.141 The sample sizes are both large enough to use the normal distribution. For a confidence interval using the normal distribution, we use

$$\text{Sample statistic} \pm z^* \cdot SE.$$

The relevant sample statistic for a confidence interval for a difference in proportions is $\hat{p}_1 - \hat{p}_2 = 114/150 - 135/150 = 0.76 - 0.90$. For a 99% confidence interval, we have $z^* = 2.576$, and we use the sample proportions in computing the standard error. The confidence interval is

$$
\begin{aligned}
(\hat{p}_1 - \hat{p}_2) \quad &\pm \quad z * \cdot \sqrt{\frac{\hat{p}_1(1-\hat{p}_1)}{n_1} + \frac{\hat{p}_2(1-\hat{p}_2)}{n_2}} \\
(0.76 - 0.90) \quad &\pm \quad 2.576 \cdot \sqrt{\frac{0.76(0.24)}{150} + \frac{0.90(0.10)}{150}} \\
-0.14 \quad &\pm \quad 0.110 \\
-0.25 \quad &\text{to} \quad -0.03
\end{aligned}
$$

The best estimate for the difference in the two proportions $p_1 - p_2$ is -0.14, the margin of error is ± 0.11, and the 99% confidence interval for $p_1 - p_2$ is -0.25 to -0.03.

6.143 Using I for the internet users and N for the non-internet users, we see that

$$\hat{p}_I = \frac{807}{1754} = 0.46 \qquad \text{and} \qquad \hat{p}_N = \frac{130}{483} = 0.27$$

In the sample, the internet users are more trusting. We estimate the difference in proportions $p_I - p_N$. The relevant sample statistic to estimate this difference is $\hat{p}_I - \hat{p}_N = 0.46 - 0.27$. For a 90% confidence interval, we have $z^* = 1.645$. The confidence interval is

$$\text{Sample statistic} \quad \pm \quad z^* \cdot SE$$

$$(\hat{p}_I - \hat{p}_N) \quad \pm \quad z* \cdot \sqrt{\frac{\hat{p}_I(1 - \hat{p}_I)}{n_I} + \frac{\hat{p}_N(1 - \hat{p}_N)}{n_N}}$$

$$(0.46 - 0.27) \quad \pm \quad 1.645 \cdot \sqrt{\frac{0.46(0.54)}{1754} + \frac{0.27(0.73)}{483}}$$

$$0.19 \quad \pm \quad 0.039$$

$$0.151 \quad \text{to} \quad 0.229$$

We are 90% confident that the proportion of internet users who agree that most people can be trusted is between 0.151 and 0.229 higher than the proportion of people who do not use the internet who agree with that statement.

6.145 Letting $\hat{p}_e$ and $\hat{p}_w$ represent the proportion of errors in electronic and written prescriptions, respectively, we have

$$\hat{p}_e = \frac{254}{3848} = 0.066 \qquad \text{and} \qquad \hat{p}_w = \frac{1478}{3848} = 0.384$$

The sample sizes are both very large, so it is reasonable to use a normal distribution. For 95% confidence the standard normal endpoint is $z^* = 1.96$. This gives

$$(\hat{p}_e - \hat{p}_w) \quad \pm \quad z^* \cdot \sqrt{\frac{\hat{p}_e(1 - \hat{p}_e)}{n_e} + \frac{\hat{p}_w(1 - \hat{p}_w)}{n_w}}$$

$$(0.066 - 0.384) \quad \pm \quad 1.96 \cdot \sqrt{\frac{0.066(1 - 0.066)}{3848} + \frac{0.384(1 - 0.384)}{3848}}$$

$$-0.318 \quad \pm \quad 0.017$$

$$-0.335 \quad \text{to} \quad -0.301$$

The margin of error is very small because the sample size is so large. Note that if we had subtracted the other way to find a confidence interval for $p_w - p_e$, the interval would be 0.301 to 0.335, with the same interpretation. In each case, we are 95% sure that the error rate is between 0.335 and 0.301 less for electronic prescriptions. This is a very big change! Since zero is not in this interval, it is not plausible that there is no difference. We can be confident that there are fewer errors with electronic prescriptions.

6.147 We have $\hat{p}_C = 421/832 = 0.506$ and $\hat{p}_N = 810/1920 = 0.422$. The sample sizes are both very large, so it is reasonable to use a normal distribution. For 90% confidence the standard normal endpoint is $z^* = 1.645$. This gives

$$\text{Statistic} \quad \pm \quad z^* \cdot SE$$

$$(\hat{p}_C - \hat{p}_N) \quad \pm \quad z^* \cdot \sqrt{\frac{\hat{p}_C(1 - \hat{p}_C)}{n_C} + \frac{\hat{p}_N(1 - \hat{p}_N)}{n_N}}$$

$$(0.506 - 0.422) \quad \pm \quad 1.645 \cdot \sqrt{\frac{0.506(1 - 0.506)}{832} + \frac{0.422(1 - 0.422)}{1920}}$$

$$0.084 \quad \pm \quad 0.034$$

$$0.050 \quad \text{to} \quad 0.118$$

We are 90% sure that the proportion of people with children visiting the public library is between 0.050 and 0.118 higher than the proportion of people without children visiting the public library.

6.149 We have $\hat{p}_M = 0.32$ and $\hat{p}_E = 0.44$. The 95% confidence interval is

$$
\begin{aligned}
(\hat{p}_M - \hat{p}_E) \quad &\pm \quad z^* \cdot \sqrt{\frac{\hat{p}_M(1 - \hat{p}_M)}{n_M} + \frac{\hat{p}_E(1 - \hat{p}_E)}{n_E}} \\
(0.32 - 0.44) \quad &\pm \quad 1.96 \cdot \sqrt{\frac{0.32(0.68)}{122} + \frac{0.44(0.56)}{160}} \\
-0.12 \quad &\pm \quad 0.11 \\
-0.23 \quad &\text{to} \quad -0.01
\end{aligned}
$$

We are 95% sure that breeding success is between 0.23 and 0.01 *less* for metal tagged penguins. This shows a significant difference at a 5% level.

6.151 Using *StatKey* or other technology to create a bootstrap distribution of the differences in sample proportions, we see for one set of 1000 simulations that $SE = 0.048$. (Answers may vary slightly with other simulations.)

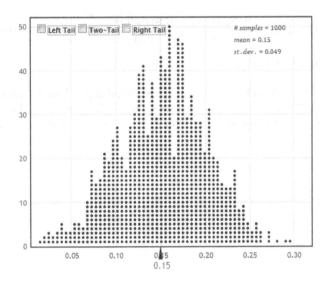

Using the formula with $\hat{p}_A = 90/120 = 0.75$ and $\hat{p}_B = 180/300 = 0.60$ as estimates for the two population proportions, we have

$$
SE = \sqrt{\frac{p_A(1 - p_A)}{n_A} + \frac{p_B(1 - p_B)}{n_B}} \approx \sqrt{\frac{0.75(0.25)}{120} + \frac{0.60(0.40)}{300}} = 0.049
$$

We see that the bootstrap standard error and the formula match very closely.

6.153 We use *StatKey* or other technology to create a bootstrap distribution with at least 1000 simulated differences in proportion. We find the endpoints that contain 95% of the simulated statistics and see that this 95% confidence interval is 0.160 to 0.266.

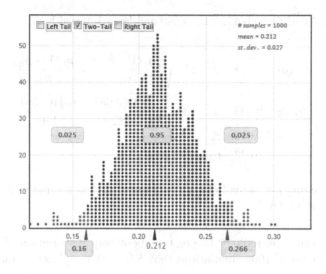

Using the normal distribution and the formula for standard error, we have

$$(0.82 - 0.61) \pm 1.96 \cdot \sqrt{\frac{0.82(0.18)}{460} + \frac{0.61(0.39)}{520}} = 0.21 \pm 0.055 = (0.155, 0.265).$$

The two methods give very similar results.

6.155 We find the proportion who die for males $\hat{p}_m$ and females $\hat{p}_f$. Using technology, we have $\hat{p}_m = 0.194$ with $n_m = 124$ and $\hat{p}_f = 0.211$ with $n_f = 76$. Also using technology, we find the 95% confidence interval for $p_m - p_f$ to be -0.132 to 0.098. The proportion who die is between 0.132 lower and 0.098 higher for males than it is for females.

Section 6.3-HT Solutions

6.157   (a) For Group A, the proportion who survive is $\hat{p}_A = 63/82 = 0.768$. For Group B, the proportion who survive is $\hat{p}_B = 31/67 = 0.463$. For the pooled proportion, we combine the two groups and look at the overall proportion who survived. The combined group has $82 + 67 = 149$ people in it, and $63 + 31 = 94$ of these people survived, so the pooled proportion is $\hat{p} = 94/149 = 0.631$.

  (b) This is a test for a difference in proportions, and we have $H_0 : p_A = p_B$ vs $H_a : p_A > p_B$. The sample sizes are large enough to use the normal distribution. In general, the standardized test statistic is

$$z = \frac{\text{Sample Statistic} - \text{Null Parameter}}{SE}.$$

In this test for a difference in proportions, the sample statistic is $\hat{p}_A - \hat{p}_B$ and the parameter from the null hypothesis is 0, since we have $H_0 : p_A - p_B = 0$. The standard error uses the pooled proportion. The standardized test statistic is

$$z = \frac{\hat{p}_A - \hat{p}_B}{\sqrt{\frac{\hat{p}(1-\hat{p})}{n_A} + \frac{\hat{p}(1-\hat{p})}{n_B}}} = \frac{0.768 - 0.463}{\sqrt{\frac{0.631(0.369)}{82} + \frac{0.631(0.369)}{67}}} = 3.84.$$

This is an upper-tail test, so the p-value is the area above 3.84 in a standard normal distribution. We see that the p-value is essentially zero. The p-value is very small so we find strong evidence that Treatment A is significantly better.

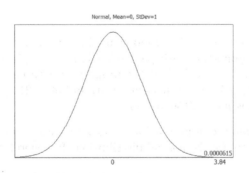

6.159   (a) For males, the proportion who plan to vote yes is $\hat{p}_m = 0.24$. For females, the proportion who plan to vote yes is $\hat{p}_f = 0.32$. For the pooled proportion, we need to know not just the proportions, but how many males and females in the samples plan to vote yes. We see that $0.24 \cdot 50 = 12$ males plan to vote yes and $0.32 \cdot 50 = 16$ females plan to vote yes in the samples. We combine the two groups and look at the proportion who plan to vote yes. The combined group has $50 + 50 = 100$ people in it, and $12 + 16 = 28$ of them plan to vote yes, so the pooled proportion is $\hat{p} = 28/100 = 0.28$.

  (b) This is a test for a difference in proportions, and we have $H_0 : p_m = p_f$ vs $H_a : p_m < p_f$. The sample sizes are (just barely) large enough to use the normal distribution. In general, the standardized test statistic is

$$z = \frac{\text{Sample Statistic} - \text{Null Parameter}}{SE}.$$

In this test for a difference in proportions, the sample statistic is $\hat{p}_m - \hat{p}_f$ and the parameter from the null hypothesis is 0, since we have $H_0 : p_m - p_f = 0$. The standard error uses the pooled proportion. The standardized test statistic is

$$z = \frac{\hat{p}_m - \hat{p}_f}{\sqrt{\frac{\hat{p}(1-\hat{p})}{n_m} + \frac{\hat{p}(1-\hat{p})}{n_f}}} = \frac{0.24 - 0.32}{\sqrt{\frac{0.28(0.72)}{50} + \frac{0.28(0.72)}{50}}} = -0.89.$$

This is a lower-tail test, so the p-value is the area below -0.89 in a standard normal distribution. Using technology or a table, we see that the p-value is 0.187. This p-value is quite large so we do not find evidence of any difference between males and females in the proportion who support the initiative. Note that it is likely that this lack of evidence is partly due to the small sample sizes. Further investigation with larger sample sizes might likely show evidence of a difference.

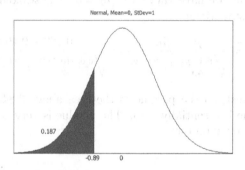

6.161   (a) For the treatment group, the proportion with pain relief is $\hat{p}_T = 36/75 = 0.48$. For the control (placebo) group, the proportion with pain relief is $\hat{p}_C = 21/75 = 0.28$. For the pooled proportion, we combine the two groups and look at the proportion with pain relief for the combined group. The combined group has $75 + 75 = 150$ patients in it, and $36 + 21 = 57$ of them had some pain relief, so the pooled proportion is $\hat{p} = 57/150 = 0.38$.

(b) This is a test for a difference in proportions, and we have $H_0 : p_T = p_C$ vs $H_a : p_T > p_C$. The sample sizes are large enough to use the normal distribution. In general, the standardized test statistic is

$$z = \frac{\text{Sample Statistic} - \text{Null Parameter}}{SE}.$$

In this test for a difference in proportions, the sample statistic is $\hat{p}_T - \hat{p}_C$ and the parameter from the null hypothesis is 0, since we have $H_0 : p_T - p_C = 0$. The standard error uses the pooled proportion. The standardized test statistic is

$$z = \frac{\hat{p}_T - \hat{p}_C}{\sqrt{\frac{\hat{p}(1-\hat{p})}{n_T} + \frac{\hat{p}(1-\hat{p})}{n_C}}} = \frac{0.48 - 0.28}{\sqrt{\frac{0.38(0.62)}{75} + \frac{0.38(0.62)}{75}}} = 2.52.$$

This is an upper-tail test, so the p-value is the area above 2.52 in a standard normal distribution. Using technology or a table, we see that the p-value is 0.006. The p-value is very small, and gives strong evidence that the treatment is better than a placebo at relieving pain.

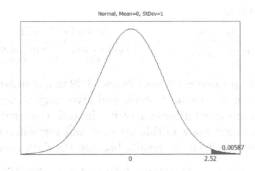

6.163 This is a test for a difference in proportions, and we define p_F and p_N to be the proportion of men copying their partners sentence structure with a fertile partner and non-fertile partner, respectively. The hypotheses are:

$$H_0: \quad p_F = p_N$$
$$H_a: \quad p_F < p_N$$

We compute the sample proportions and the pooled sample proportion:

$$\hat{p}_F = \frac{30}{62} = 0.484 \qquad \hat{p}_N = \frac{38}{61} = 0.623 \qquad \hat{p} = \frac{68}{123} = 0.553$$

It is appropriate to use a normal distribution so we compute the standardized test statistic:

$$\frac{\text{Sample statistic} - \text{Null parameter}}{SE} = \frac{(0.484 - 0.623) - 0}{\sqrt{\frac{0.553(0.447)}{62} + \frac{0.553(0.447)}{61}}} = -1.55.$$

This is a lower-tail test so the p-value is the area below -1.55 in a standard normal distribution. We see that the p-value is 0.0606. This p-value is small enough to be significant at the 10% level but not quite at the 5% level. We do not reject H_0 and do not find evidence that ovulating women affect men's speech. (However, the results are so close to being significant that it is probably worth replicating the experiment with a larger sample size.)

6.165 This is a test for a difference in proportions. Using T for the treatment group (taking the daily low-dose aspirin) and C for the control group (taking a placebo), the hypotheses are $H_0 : p_T = p_C$ vs $H_a : p_T < p_C$. The sample sizes are very large (large enough to get more than 10 heart attacks in each group), so we can use the normal distribution. In general, the standardized test statistic is

$$z = \frac{\text{Sample Statistic} - \text{Null Parameter}}{SE}.$$

In this test for a difference in proportions, the sample statistic is $\hat{p}_T - \hat{p}_C$ and the parameter from the null hypothesis is 0, since we have $H_0 : p_T - p_C = 0$. The standard error uses the pooled proportion. The three relevant proportions are:

$$\hat{p}_T = \frac{104}{11,037} = 0.0094$$
$$\hat{p}_C = \frac{189}{11,034} = 0.0171$$
$$\text{Pooled proportion} = \hat{p} = \frac{104 + 189}{11,037 + 11,034} = \frac{293}{22,071} = 0.0133$$

The standardized test statistic is

$$z = \frac{\hat{p}_T - \hat{p}_C}{\sqrt{\frac{\hat{p}(1-\hat{p})}{n_T} + \frac{\hat{p}(1-\hat{p})}{n_C}}} = \frac{0.0094 - 0.0171}{\sqrt{\frac{0.0133(0.9867)}{11,037} + \frac{0.0133(0.9867)}{11,034}}} = -4.99.$$

This is a lower-tail test, so the p-value is the area below -4.99 in a standard normal distribution. The p-value is essentially zero. The p-value is extremely small, and gives very strong evidence that taking a daily low-dose aspirin reduces the risk of having a heart attack. (Indeed, the results are so strong that the study was stopped early in order to let others know of this low-cost and very effective treatment, used broadly today.) We can infer a causal relationship from the results because the data come from a randomized experiment.

6.167 The sample statistics are $\hat{p}_o = 345/500 = 0.69$ and $\hat{p}_c = 320/500 = 0.64$. The pooled proportion is $\hat{p} = 665/1000 = 0.665$. The standardized test statistic is

$$z = \frac{\text{Statistic} - \text{Null value}}{SE} = \frac{(0.69 - 0.64) - 0}{\sqrt{\frac{0.665(0.335)}{500} + \frac{0.665(0.335)}{500}}} = 1.675$$

We find the p-value as the proportion of the standard normal distribution beyond 1.675 in the right tail, yielding a p-value of 0.047. This p-value is just below the 5% level, so we can reject H_0. After 15 days, we have enough evidence to conclude that the proportion alive for fruit flies that eat organic bananas is higher than for fruit flies that eat conventional bananas.

6.169 The sample statistics are $\hat{p}_o = 250/500 = 0.50$ and $\hat{p}_c = 130/500 = 0.260$. The pooled proportion is $\hat{p} = 380/1000 = 0.380$. The standardized test statistic is

$$z = \frac{\text{Statistic} - \text{Null value}}{SE} = \frac{(0.50 - 0.26) - 0}{\sqrt{\frac{0.38(0.62)}{500} + \frac{0.38(0.62)}{500}}} = 7.818.$$

We find the p-value as the proportion of the standard normal distribution beyond 7.818 in the right tail, yielding a p-value of 0.000. (Recall that the z test statistic is a z-score, and a z-score of 7.818 is very far out in the tail! We don't even really need a standard normal distribution to know that the p-value here is going to be very close to zero.) We reject H_0 and find very strong evidence that, after 20 days, the proportion alive of fruit flies that eat organic potatoes is significantly higher than the proportion alive that eat conventional potatoes.

6.171 Since the number in the control group that solved the problem is only 4 (which is definitely less than 10) and only 8 in the electrical stimulation group did not solve it, the sample size is too small to use the normal distribution for this test. A randomization test is more appropriate for this data.

6.173   (a) Let p_o and p_u be the proportion of Oocyst and unexposed mosquitoes, respectively that approach a human. The sample statistic is

$$\hat{p}_o - \hat{p}_u = \frac{20}{113} - \frac{36}{117} = 0.177 - 0.308 = -0.131.$$

The pooled proportion is

$$\hat{p} = \frac{20 + 36}{113 + 117} = 0.2435.$$

The standard error is

$$SE = \sqrt{\frac{\hat{p}(1-\hat{p})}{n_1} + \frac{\hat{p}(1-\hat{p})}{n_2}} = \sqrt{\frac{0.243(1-0.243)}{113} + \frac{0.243(1-0.243)}{117}} = 0.057.$$

The z-statistic then is

$$z = \frac{statistic - null}{SE} = \frac{-0.131 - 0}{0.057} = -2.30.$$

(b) Since we are testing $H_0 : p_o = p_u$ vs $H_a : p_o < p_u$, we look in the lower tail of a standard normal distribution below $z = -2.3$ to find the p-value of 0.011.

(c) Let p_s and p_u be the proportion of Sporozoite and unexposed mosquitoes, respectively that approach a human. The sample statistic is

$$\hat{p}_s - \hat{p}_u = \frac{37}{149} - \frac{14}{144} = 0.248 - 0.097 = 0.151.$$

The pooled proportion is

$$\hat{p} = \frac{37 + 14}{149 + 144} = 0.174.$$

The standard error is

$$SE = \sqrt{\frac{\hat{p}(1-\hat{p})}{n_1} + \frac{\hat{p}(1-\hat{p})}{n_2}} = \sqrt{\frac{0.174(1-0.174)}{149} + \frac{0.174(1-0.174)}{144}} = 0.044.$$

The z-statistic then is

$$z = \frac{statistic - null}{SE} = \frac{0.151 - 0}{0.044} = 3.43.$$

(d) Since we are testing $H_0 : p_s = p_u$ vs $H_a : p_s > p_u$, we look in the upper tail of a standard normal distribution above $z = 3.43$ to find the p-value of 0.0003.

(e) We have enough evidence to conclude that in the Oocyst stage mosquitoes exposed to malaria approach the human less often than mosquitoes not exposed to malaria and in the Sporozoite stage mosquitoes exposed to malaria approach the human more often than mosquitoes not exposed to malaria.

(f) Yes, we can conclude that being exposed to malaria (as opposed to not being exposed to malaria) *causes* these behavior changes in mosquitoes, because this was a randomized experiment (it was randomly determined which mosquitoes ate from the mouse infected with malaria and which mosquitoes ate from the non-infected mouse).

6.175 These are large sample sizes so the normal distribution is appropriate. The sample proportions are $\hat{p}_H = 166/8506 = 0.0195$ and $\hat{p}_P = 124/8102 = 0.0153$ in the HRT group and the placebo group, respectively. The pooled proportion is

$$\hat{p} = \frac{166 + 124}{8506 + 8102} = 0.0175.$$

The test statistic is

$$z = \frac{0.0195 - 0.0153}{\sqrt{\frac{0.0175(1-0.0175)}{8506} + \frac{0.0175(1-0.0175)}{8102}}} = \frac{0.0042}{0.002} = 2.07$$

Answers may vary slightly due to roundoff. The area in the upper tail of the standard normal distribution is 0.019, which we double to find the p-value of 0.038. Because this was a randomized experiment, we can conclude causality. There is evidence that HRT significantly increases risk of invasive breast cancer.

6.177 These are large sample sizes so the normal distribution is appropriate. The sample proportions are $\hat{p}_H = 650/8506 = 0.076$ and $\hat{p}_P = 788/8102 = 0.097$ for the HRT group and the placebo group, respectively. The pooled proportion is

$$\hat{p} = \frac{650 + 788}{8506 + 8102} = 0.087.$$

The test statistic is

$$z = \frac{0.076 - 0.097}{\sqrt{\frac{0.087(1-0.087)}{8506} + \frac{0.087(1-0.087)}{8102}}} = \frac{-0.021}{0.0044} = -4.77.$$

The area in the lower tail of the standard normal distribution below -4.77 is very small, so even after doubling we have p-value ≈ 0. Because this was a randomized experiment, we can conclude causality. There is strong evidence that HRT significantly decreases risk of fractures.

6.179 The hypotheses are $H_0 : p_m = p_f$ vs $H_a : p_m \neq p_f$ where are p_m and p_f are the proportion having surgery of males and females, respectively. Using technology, we see $\hat{p}_m = 0.565$ with $n_m = 124$ and $\hat{p}_f = 0.487$ with $n_f = 76$. Also using technology, we find that the standardized z-statistic is $z = 1.07$ and the p-value for this two-tailed test is 0.285. There is not sufficient evidence to find a difference between males and females in the proportion having surgery.

Section 6.4-D Solutions

6.181 The differences in sample means will have a standard error of

$$SE = \sqrt{\frac{\sigma_1^2}{n_1} + \frac{\sigma_2^2}{n_2}} = \sqrt{\frac{3.7^2}{25} + \frac{7.6^2}{40}} = 1.41$$

6.183 The differences in sample means will have a standard error of

$$SE = \sqrt{\frac{\sigma_1^2}{n_1} + \frac{\sigma_2^2}{n_2}} = \sqrt{\frac{18^2}{300} + \frac{22^2}{500}} = 1.43$$

6.185 We use the smaller sample size and subtract 1 to find the degrees of freedom, so we use a t-distribution with df = 7. We see that the values with 5% beyond them in each tail are ±1.89.

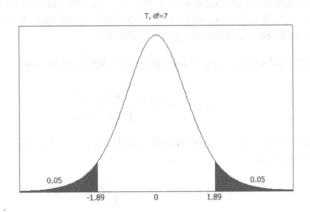

6.187 We use a t-distribution with df = 11. We see that the probability the t-statistic is greater than 2.1 is 0.0298. (With a paper table, we may only be able to specify that the area is between 0.025 and 0.05.)

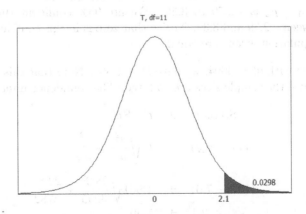

Section 6.4-CI Solutions

6.189 For a confidence interval for $\mu_1 - \mu_2$ using the t-distribution, we use

$$(\overline{x}_1 - \overline{x}_2) \pm t^* \sqrt{\frac{s_1^2}{n_1} + \frac{s_2^2}{n_2}}$$

The sample sizes are both 50, so we use a t-distribution with df = 49. For a 90% confidence interval, we have $t^* = 1.68$. The confidence interval is

$$
\begin{aligned}
(10.1 - 12.4) \quad &\pm \quad 1.68 \cdot \sqrt{\frac{2.3^2}{50} + \frac{5.7^2}{50}} \\
-2.3 \quad &\pm \quad 1.46 \\
-3.76 \quad &\text{to} \quad -0.84
\end{aligned}
$$

The best estimate for the difference in means $\mu_1 - \mu_2$ is -2.3, the margin of error is ± 1.46, and the 90% confidence interval for $\mu_1 - \mu_2$ is -3.76 to -0.84. We are 90% confident that the mean for population 2 is between 0.84 and 3.76 more than the mean for population 1.

6.191 For a confidence interval for $\mu_1 - \mu_2$ using the t-distribution, we use

$$(\overline{x}_1 - \overline{x}_2) \pm t^* \sqrt{\frac{s_1^2}{n_1} + \frac{s_2^2}{n_2}}$$

The smaller sample size is 8, so we use a t-distribution with df = 7. For a 95% confidence interval, we have $t^* = 2.36$. The confidence interval is

$$
\begin{aligned}
(5.2 - 4.9) \quad &\pm \quad 2.36 \cdot \sqrt{\frac{2.7^2}{10} + \frac{2.8^2}{8}} \\
0.3 \quad &\pm \quad 3.09 \\
-2.79 \quad &\text{to} \quad 3.39
\end{aligned}
$$

The best estimate for the difference in means $\mu_1 - \mu_2$ is 0.3, the margin of error is ± 3.09, and the 95% confidence interval for $\mu_1 - \mu_2$ is -2.79 to 3.39. We are 95% confident that the difference in the two population means is between -2.79 and 3.39. Notice that zero is in this confidence interval, so it is certainly possible that the two population means are equal.

6.193 For 95% confidence with df = 3094, we have $t^* = 1.961$. Note that this is very close to the standard normal value of 1.960, since the sample sizes are so large. The confidence interval is:

$$
\begin{aligned}
\text{Statistic} \quad &\pm \quad t^* \cdot SE \\
(\overline{x}_F - \overline{x}_N) \quad &\pm \quad t^* \sqrt{\frac{s_F^2}{n_F} + \frac{s_N^2}{n_N}} \\
(10.1 - 7.0) \quad &\pm \quad 1.961 \sqrt{\frac{38.9^2}{3095} + \frac{22.8^2}{5782}} \\
3.1 \quad &\pm \quad 1.49 \\
1.61 \quad &\text{to} \quad 4.59
\end{aligned}
$$

We are 95% sure that the mean concentration of DiNP in people who have recently eaten fast food is between 1.6 ng/mL and 4.6 ng/mL higher than in people who have not recently eaten fast food.

6.195 (a) *Randomized* means the participants were divided randomly between the two groups. *Double-blind* means neither the participants nor those measuring vascular health knows which group participants are in. *Placebo-controlled* means participants are given a treatment that is almost identical to the real treatment – in this case, dark chocolate which has had the flavonoids removed.

(b) We are estimating $\mu_C - \mu_N$ where μ_C represents the mean increase in flow-mediated dilation for people eating dark chocolate every day and μ_N represents the mean increase in flow-mediated dilation for people eating a dark chocolate substitute each day. For a 95% confidence interval using a t-distribution with degrees of freedom equal to 9, we use $t^* = 2.26$. We have:

$$\text{Sample statistic} \quad \pm \quad t^* \cdot SE$$

$$(\overline{x}_C - \overline{x}_N) \quad \pm \quad 2.26\sqrt{\frac{s_C^2}{n_C} + \frac{s_N^2}{n_N}}$$

$$(1.3 - (-0.96)) \quad \pm \quad 2.26\sqrt{\frac{2.32^2}{11} + \frac{1.58^2}{10}}$$

$$2.26 \quad \pm \quad 1.94$$

$$0.32 \quad \text{to} \quad 4.20$$

We are 95% confident that the mean increase in flow-mediated dilation for those in the dark chocolate group is between 0.32 and 4.20 higher than for those in the fake chocolate group.

(c) No. "No difference" implies that $\mu_C - \mu_N = 0$. We see that 0 is not within the confidence interval of 0.32 to 4.20 and, in fact, all plausible values show a positive difference.

6.197 Since both samples have size 24, we use a t-distribution with $24 - 1 = 23$ degrees of freedom to find a 95% point of $t^* = 2.069$.

$$\overline{x}_I - \overline{x}_S \quad \pm \quad t^*\sqrt{\frac{s_I^2}{n_I} + \frac{s_S^2}{n_S}}$$

$$37.29 - 50.92 \quad \pm \quad 2.069\sqrt{\frac{12.54^2}{24} + \frac{14.33^2}{24}}$$

$$-13.63 \quad \pm \quad 8.04$$

$$-21.67 \quad \text{to} \quad -5.59$$

We are 95% sure that diners using individual bills at this restaurant will spend, on average, somewhere between 5.59 and 21.67 shekels less than diners who are splitting the bill

6.199 We let μ_N and μ_Y represent mean GPA for students whose roommate does not bring a videogame (No) or does bring one (Yes) to campus, respectively. The two relevant sample sizes are 44 and 40, so the smallest sample size in these two groups is 40 and the degrees of freedom are $df = 39$. For a 95% confidence level with $df = 39$, we have $t^* = 2.02$. Calculating the 95% confidence interval for the difference in means:

$$\text{Sample statistic} \quad \pm \quad t^* \cdot SE$$

$$(\overline{x}_N - \overline{x}_Y) \quad \pm \quad t^* \sqrt{\frac{s_N^2}{n_N} + \frac{s_Y^2}{n_Y}}$$

$$(3.039 - 2.754) \quad \pm \quad 2.02\sqrt{\frac{0.689^2}{44} + \frac{0.639^2}{40}}$$

$$0.285 \quad \pm \quad 0.293$$

$$-0.008 \quad \text{to} \quad 0.578$$

We are 95% sure that, for students who do bring a videogame to campus, mean GPA for students whose roommate does not bring a videogame will be between 0.008 lower and 0.578 higher than the mean for students whose roommate does bring a videogame. Since 0 is (just barely!) in this interval, it is possible that there is no effect from the roommate bringing a videogame.

6.201 We let μ_N and μ_Y represent mean GPA for students who do not bring a videogame (No) or do bring one (Yes) to campus, respectively. The two relevant sample sizes are 38 and 40, so the smallest sample size in these two groups is 38 and the degrees of freedom are $df = 37$. For a 95% confidence level with $df = 37$, we have $t^* = 2.03$. Calculating the 95% confidence interval for the difference in means:

$$\text{Sample statistic} \quad \pm \quad t^* \cdot SE$$

$$(\overline{x}_N - \overline{x}_Y) \quad \pm \quad t^* \sqrt{\frac{s_N^2}{n_N} + \frac{s_Y^2}{n_Y}}$$

$$(2.932 - 2.754) \quad \pm \quad 2.03\sqrt{\frac{0.699^2}{38} + \frac{0.639^2}{40}}$$

$$0.178 \quad \pm \quad 0.308$$

$$-0.130 \quad \text{to} \quad 0.486$$

We are 95% sure that, for students whose roommate brings a videogame to campus, mean GPA for students who do not bring a videogame will be between 0.130 lower and 0.486 higher than the mean for students who bring a videogame. Since 0 is in this interval, it is possible that there is no effect from bringing a videogame.

6.203 We use *StatKey* or other technology to create a bootstrap distribution. We see for one set of 1000 simulations that $SE \approx 1.12$. (Answers may vary slightly with other simulations.)

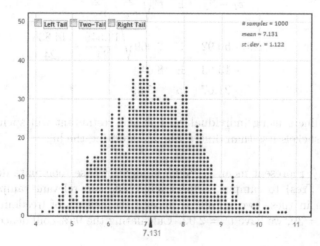

Using the formula from the Central Limit Theorem, and using the sample standard deviations as estimates of the population standard deviations, we have

$$SE = \sqrt{\frac{s_1^2}{n_1} + \frac{s_2^2}{n_2}} = \sqrt{\frac{20.72^2}{500} + \frac{14.23^2}{500}} = 1.124.$$

We see that the bootstrap standard error and the formula match very closely.

6.205 (a) We see that there are 168 females and 193 males.

(b) The males in the sample exercise more, with a mean of 9.876 compared to the female mean of 8.110. The difference is $9.876 - 8.110 = 1.766$ hours per week more exercise for the males, on average.

(c) We are estimating $\mu_f - \mu_m$, where μ_f represents the number of hours spent exercising per week by all female students at this university and μ_m represents the number of hours spent exercising a week for all male students at this university. For a 95% confidence interval with degrees of freedom 167, we use $t^* = 1.97$. The confidence interval is

$$\text{Sample statistic} \quad \pm \quad t^* \cdot SE$$
$$(\overline{x}_f - \overline{x}_m) \quad \pm \quad 1.97\sqrt{\frac{s_f^2}{n_f} + \frac{s_m^2}{n_m}}$$
$$(8.110 - 9.876) \quad \pm \quad 1.97\sqrt{\frac{5.199^2}{168} + \frac{6.069^2}{193}}$$
$$-1.766 \quad \pm \quad 1.168$$
$$-2.934 \quad \text{to} \quad -0.598$$

A 95% confidence interval for the difference in means is -2.93 to -0.60.

(d) Up to two decimal places, the confidence interval we found is the same as the one given in the computer output. The small differences are probably due to round-off error. Note you might have switched the order to estimate $\mu_m - \mu_f$ but that would only change the signs of the interval.

(e) We are 95% confident that the average amount males exercise, in hours per week, is between 0.60 and 2.93 hours more than the amount females exercise per week, for students at this university.

6.207 Using any statistics package, we see that a 95% confidence interval for $\mu_f - \mu_m$ is -2.36 to 0.91. We are 95% sure that the difference in average number of grams of fiber eaten in a day between males and females is between 2.36 more for males and 0.91 more for females. "No difference" is a plausible option since the interval contains 0.

Section 6.4-HT Solutions

6.209 In general, the standardized test statistic is

$$\frac{\text{Sample Statistic} - \text{Null Parameter}}{SE}$$

In this test for a difference in means, the sample statistic is $\overline{x}_1 - \overline{x}_2$ and the parameter from the null hypothesis is 0 since the null hypothesis statement is equivalent to $\mu_1 - \mu_2 = 0$. Substituting the formula for the standard error for a difference in two means, we compute the t-test statistic to be:

$$t = \frac{(\overline{x}_1 - \overline{x}_2) - 0}{\sqrt{\frac{s_1^2}{n_1} + \frac{s_2^2}{n_2}}} = \frac{56 - 51}{\sqrt{\frac{8.2^2}{30} + \frac{6.9^2}{40}}} = 2.70$$

This is an upper-tail test, so the p-value is the area above 2.70 in a t-distribution with df $= 29$. (Since both sample sizes are greater than 30, we could also use the normal distribution to estimate the p-value.) We see that the p-value is about 0.006. This p-value is very small, so we reject H_0 and find evidence to support the alternative hypothesis that $\mu_1 > \mu_2$.

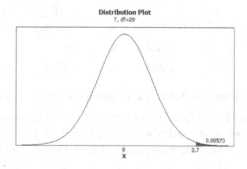

6.211 In general, the standardized test statistic is

$$\frac{\text{Sample Statistic} - \text{Null Parameter}}{SE}$$

In this test for a difference in means, the sample statistic is $\overline{x}_A - \overline{x}_B$ and the parameter from the null hypothesis is 0 since the null hypothesis statement is equivalent to $\mu_A - \mu_B = 0$. Substituting the formula for the standard error for a difference in two means, we compute the t-test statistic to be:

$$t = \frac{(\overline{x}_A - \overline{x}_B) - 0}{\sqrt{\frac{s_A^2}{n_A} + \frac{s_B^2}{n_B}}} = \frac{125 - 118}{\sqrt{\frac{18^2}{8} + \frac{14^2}{15}}} = 0.96$$

This is a two-tailed test, so the p-value is two times the area above 0.96 in a t-distribution with df $= 7$. We see that the p-value is $2(0.185) = 0.37$. This p-value is larger than any reasonable significance level, so we do not reject H_0. There is not enough evidence to support the alternative hypothesis that the means are different.

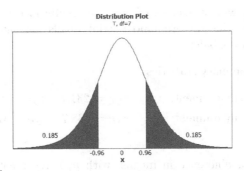

6.218 (a) Let μ_U and μ_E be the average time spent looking at the unexpected and expected populations, respectively. The hypotheses are

$$H_0 : \mu_U = \mu_E$$
$$H_a : \mu_U > \mu_E.$$

(b) The relevant sample statistic is

$$\overline{x}_U - \overline{x}_E = 9.9 - 7.5 = 2.4 \text{ seconds.}$$

(c) The t-statistic is

$$t = \frac{statistic - null}{SE} = \frac{\overline{x}_U - \overline{x}_E - 0}{\sqrt{\frac{s_U^2}{n_U} + \frac{s_E^2}{n_E}}} = \frac{9.9 - 7.5}{\sqrt{\frac{4.5^2}{20} + \frac{4.2^2}{20}}} = 1.74.$$

(d) The p-value is the proportion above $t = 1.74$ on a t-distribution with 19 degrees of freedom, which is 0.049.

(e) The p-value of 0.049 is less than $\alpha = 0.10$, so we reject H_0.

(f) We have somewhat convincing evidence that babies look longer at the unexpected population after viewing the sample data.

6.215 The hypotheses are $H_0 : \mu_m = \mu_f$ vs $H_a : \mu_m > \mu_f$, where μ_m and μ_f represent the mean salary recommended for male applicants and female applicants, respectively. The sample sizes are large enough to use the t-distribution. The t-test statistic is:

$$t = \frac{\text{Sample statistic} - \text{Null parameter}}{SE} = \frac{(\overline{x}_m - \overline{x}_f) - 0}{\sqrt{\frac{s_m^2}{n_m} + \frac{s_f^2}{n_f}}} = \frac{(30238 - 26508) - 0}{\sqrt{\frac{5152^2}{63} + \frac{7348^2}{64}}} = 3.316$$

This is a right-tail test, so the p-value is the area above 3.316 in a t-distribution with df = 62. We see that the p-value is 0.0008. This is a very small p-value, so we reject H_0, and conclude that there is very strong evidence that there is a gender bias in salary in favor of male applicants.

6.217 This test of difference in means is testing the null hypothesis that the Patriots average air pressure and Colts average air pressure was the same ($H_0 : \mu_p = \mu_c$), against the alternative that the Patriots average air pressure was less than the Colts ($H_0 : \mu_p < \mu_c$). The test statistic is

$$t = \frac{\overline{x}_p - \overline{x}_c}{\sqrt{\frac{s_p^2}{n_p} + \frac{s_c^2}{n_c}}} = \frac{11.10 - 12.63}{\sqrt{\frac{0.40^2}{11} + \frac{0.12^2}{4}}} = -11.36$$

Comparing to a t-distribution with 3 degrees of freedom results in a p-value of 0.00073. So we conclude that the average air pressure of the New England Patriot's balls was significantly less than the average air pressure of the Indianapolis Colt's balls.

6.219 We first compute the summary statistics

$$\text{Enriched environment:} \quad \overline{x}_{EE} = 231.7 \quad s_{EE} = 71.2 \quad n_{EE} = 7$$
$$\text{Standard environment:} \quad \overline{x}_{SE} = 438.7 \quad s_{SE} = 37.7 \quad n_{SE} = 7$$

This is a hypothesis test for a difference in means, with μ_{EE} representing the mean number of seconds spent in darkness by mice who have lived in an enriched environment and then been exposed to stress-inducing events, while μ_{SE} represents the same quantity for mice who lived in a standard environment. The hypotheses are:

$$H_0: \quad \mu_{EE} = \mu_{SE}$$
$$H_a: \quad \mu_{EE} < \mu_{SE}$$

The relevant statistic for this test is $\overline{x}_{EE} - \overline{x}_{SE}$, and the relevant null parameter is zero, since from the null hypothesis we have $\mu_{EE} - \mu_{SE} = 0$. The t-test statistic is:

$$t = \frac{\text{Sample statistic} - \text{Null parameter}}{SE} = \frac{(\overline{x}_{EE} - \overline{x}_{SE}) - 0}{\sqrt{\frac{s_{EE}^2}{n_{EE}} + \frac{s_{SE}^2}{n_{SE}}}} = \frac{231.7 - 438.7}{\sqrt{\frac{71.2^2}{7} + \frac{37.7^2}{7}}} = -6.80.$$

This is a lower-tail test, so the p-value is the area to the left of -6.80 in a t-distribution with df = 6. We see that the p-value is 0.00025. This is an extremely small p-value, so we reject H_0, and conclude that there is very strong evidence that mice who have been able to exercise in an enriched environment are better able to handle stress.

6.221 (a) This is an experiment since the subjects are randomly assigned to drink tea or coffee.

(b) This is a difference in means test and we are testing

$$H_0: \quad \mu_T = \mu_C$$
$$H_a: \quad \mu_T > \mu_C$$

where μ_T and μ_C are the mean production levels of interferon gamma in tea drinkers and coffee drinkers, respectively.

(c) For the tea drinkers, we find that $\overline{x}_T = 34.82$ with $s_T = 21.1$ and $n_T = 11$. For the coffee drinkers, we have $\overline{x}_C = 17.70$ with $s_C = 16.7$ and $n_C = 10$. The appropriate statistic is $\overline{x}_T - \overline{x}_C$ and the null parameter is zero, so we have

$$
\begin{aligned}
t &= \frac{\text{Sample statistic} - \text{Null parameter}}{SE} \\
&= \frac{(\overline{x}_T - \overline{x}_C) - 0}{\sqrt{\frac{s_T^2}{n_T} + \frac{s_C^2}{n_c}}} \\
&= \frac{34.82 - 17.70}{\sqrt{\frac{21.1^2}{11} + \frac{16.7^2}{10}}} \\
&= 2.07
\end{aligned}
$$

The smaller sample size is 10, so we find the p-value using a t-distribution with df = 9. This is an upper tail test, so we find that the p-value is 0.0342. At a 5% level, we conclude that there is evidence that mean production of this disease fighting molecule is enhanced by drinking tea.

(d) The sample sizes are small so it is hard to tell whether the data are normal or not. Dotplots for each group are shown below. Notice that most of the values are in the "tails" with few dots in the middle. This may cast some doubt on the normality condition. To be on the safe side, a randomization test might be more appropriate.

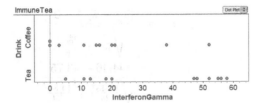

(e) To create a randomization statstic we scramble the tea/coffee assignments (so they aren't related to the interferon gamma production values) and find the difference in means between the two groups. Repeating this 1000 times produces a randomization distribution of means such as the one below.

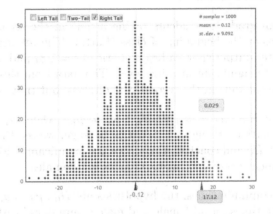

In this distribution, 29 of the 1000 randomizations produced a difference in means that was larger than the difference of 17.12 that was observed in the original sample. This gives a p-value of 0.029 which is fairly small, giving evidence that mean interferon gamma production is higher when drinking tea than when drinking coffee. Note that the p-value and strength of evidence are similar to what we found in part (c) with the t-distribution.

(f) Since this was a well-designed experiment, we can conclude that there is some evidence that drinking (lots of) tea enhances a person's immune response (at least as measured by interferon gamma production).

6.223 The hypotheses are $H_0 : \mu_M = \mu_E$ vs $H_a : \mu_M > \mu_E$, where μ_M and μ_E represent the mean arrival time (in days after November 1st) for metal tagged and electronic tagged penguins, respectively. The sample sizes are large so it is appropriate to use the t-distribution. The t-test statistic is:

$$t = \frac{\text{Sample statistic} - \text{Null parameter}}{SE} = \frac{(\overline{x}_M - \overline{x}_E) - 0}{\sqrt{\frac{s_M^2}{n_M} + \frac{s_E^2}{n_E}}} = \frac{37 - 21}{\sqrt{\frac{38.77^2}{167} + \frac{27.50^2}{189}}} = 4.44$$

This is an upper-tail test, so the p-value is the area above 4.44 in a t-distribution with df = 166. We see that the p-value is 0.000008, or essentially zero. This is an extremely small p-value, so we reject H_0, and conclude that there is very strong evidence that penguins with metal tags arrive at the breeding site later, on average, than penguins with electronic tags.

6.225 We choose a two-tailed test $H_0 : \mu_f = \mu_m$ vs $H_a : \mu_f \neq \mu_m$, where μ_f and μ_m are the mean meal costs for females and males, respectively. We compute a t-statistic

$$
\begin{aligned}
t &= \frac{(\overline{x}_f - \overline{x}_m) - 0}{\sqrt{\frac{s_f^2}{n_f} + \frac{s_m^2}{n_m}}} \\
&= \frac{44.46 - 43.75}{\sqrt{\frac{15.48^2}{24} + \frac{14.81^2}{24}}} \\
&= 0.16
\end{aligned}
$$

Using two tails from a t-distribution with $24 - 1 = 23$ degrees of freedom we find the p-value is $2 \times 0.437 = 0.874$. This is not at all a small p-value, so we have no strong evidence of a difference in the mean cost of meal orders between male and females.

6.227 Let μ_1 be the mean grade for students taking a quiz in the second half of class (Late) and μ_2 be the mean grade for quizzes at the beginning of class (Early). The relevant hypotheses are $H_0 : \mu_1 = \mu_2$ vs $H_a : \mu_1 \neq \mu_2$. The computer output shows a test statistic of $t = 1.87$ and p-value=0.066. This is a somewhat small p-value, but not quite significant at a 5% level. The data from this sample show some evidence that it might be better to have the quiz in the second part of class, but that evidence is not very strong.

6.229 The hypotheses are $H_0 : \mu_f = \mu_m$ vs $H_a : \mu_f \neq \mu_m$, where μ_f and μ_m are the respective mean exercise times in the population. Using any statistics package, we see that the t-statistic is -2.98 and the p-value is 0.003. We reject H_0 and conclude that there is a significant evidence showing a difference in mean number of hours a week spent exercising between males and females.

6.231 To compare mean commute times, the hypothesis are $H_0 : \mu_f = \mu_m$ vs $H_a : \mu_f \neq \mu_m$, where μ_f and μ_m are the mean commute times for all female and male commuters in Atlanta.

Here are summary statistics for the commute times, broken down by sex, for the data in **CommuteAtlanta**.

```
Sex   N   Mean   StDev
F    246  26.8   17.3
M    254  31.3   23.4
```

The value of the t-statistic is $t = \dfrac{26.8 - 31.3}{\sqrt{\frac{17.3^2}{246} + \frac{23.4^2}{254}}} = -2.45$

We find the p-value using a t-distribution with $246 - 1 = 245$ degrees of freedom. The area in the lower tail of this distribution below -2.47 is 0.007. Doubling to account for two tails gives p-value = 0.014. This p-value is smaller than the significance level ($\alpha = 0.05$) so we reject the null hypothesis and conclude that the average commute time for women in Atlanta is different from (and, in fact, less than) the average commute time for men.

Section 6.5 Solutions

6.233 For a confidence interval for the average difference $\mu_d = \mu_1 - \mu_2$ using the t-distribution and paired sample data, we use

$$\overline{x}_d \pm t^* \cdot \frac{s_d}{\sqrt{n_d}}$$

We use a t-distribution with df $= 99$, so for a 90% confidence interval, we have $t^* = 1.66$. The confidence interval is

$$
\begin{array}{ccc}
\overline{x}_d & \pm & t^* \cdot \dfrac{s_d}{\sqrt{n_d}} \\[2mm]
556.9 & \pm & 1.66 \cdot \dfrac{143.6}{\sqrt{100}} \\[2mm]
556.9 & \pm & 23.8 \\[1mm]
533.1 & \text{to} & 580.7
\end{array}
$$

The best estimate for the mean difference $\mu_d = \mu_1 - \mu_2$ is 556.9, the margin of error is 23.8, and the 90% confidence interval for μ_d is 533.1 to 580.7. We are 90% confident that the mean for population or treatment 1 is between 533.1 and 580.7 larger than the mean for population or treatment 2.

6.235 For paired difference in means, we begin by finding the differences $d =$ Situation 1 $-$ Situation 2 for each pair. These are shown in the table below.

Case	Difference
1	-8
2	-3
3	3
4	-16
5	-7
6	10
7	-3
8	-1

The mean of these 8 differences is $\overline{x}_d = -3.13$ with standard deviation $s_d = 7.74$. For a confidence interval for the average difference $\mu_d = \mu_1 - \mu_2$ using the t-distribution and paired sample data, we use

$$\overline{x}_d \pm t^* \cdot \frac{s_d}{\sqrt{n_d}}$$

We use a t-distribution with df $= 7$, so for a 95% confidence interval, we have $t^* = 2.36$. The confidence interval is

$$
\begin{array}{ccc}
\overline{x}_d & \pm & t^* \cdot \dfrac{s_d}{\sqrt{n_d}} \\[2mm]
-3.13 & \pm & 2.36 \cdot \dfrac{7.74}{\sqrt{8}} \\[2mm]
-3.13 & \pm & 6.46 \\[1mm]
-9.59 & \text{to} & 3.33
\end{array}
$$

The best estimate for the mean difference $\mu_d = \mu_1 - \mu_2$ is -3.13, the margin of error is 6.46, and the 95% confidence interval for μ_d is -9.59 to 3.33. We are 95% confident that the difference in means $\mu_1 - \mu_2$ is between -9.59 and $+3.33$. (Notice that 0 is within this confidence interval, so based on this sample data we cannot be sure that there is a difference between the means.)

6.237 In general, the standardized test statistic is

$$\frac{\text{Sample Statistic} - \text{Null Parameter}}{SE}$$

In this test for a paired difference in means, the sample statistic is $\overline{x}_d$ and the parameter from the null hypothesis is 0 since the null hypothesis statement is equivalent to $\mu_1 - \mu_2 = \mu_d = 0$. The standard error is $s_d/\sqrt{n_d}$, and the t-test statistic is:

$$t = \frac{\overline{x}_d - 0}{\frac{s_d}{\sqrt{n_d}}} = \frac{-2.6}{\frac{4.1}{\sqrt{18}}} = -2.69$$

This is a two-tail test so, using a t-distribution with df $= 17$, we multiply the area in the tail below -2.69 by two to obtain a p-value of $2(0.008) = 0.016$. At a 5% level, we reject the null hypothesis and find evidence that the means are not the same.

6.239 For paired difference in means, we begin by finding the differences $d = $ Situation 1 $-$ Situation 2 for each pair. These are shown in the table below.

Difference	5	11	-10	25	-7	0	20	-7	6	7

The mean of these ten differences is $\overline{x}_d = 5.0$ with standard deviation $s_d = 11.6$. In general, the standardized test statistic is

$$\frac{\text{Sample Statistic} - \text{Null Parameter}}{SE}$$

In this test for a paired difference in means, the sample statistic is $\overline{x}_d$ and the parameter from the null hypothesis is 0 since the null hypothesis statement is equivalent to $\mu_1 - \mu_2 = \mu_d = 0$. The standard error is $s_d/\sqrt{n_d}$, and the t-test statistic is:

$$t = \frac{\overline{x}_d - 0}{\frac{s_d}{\sqrt{n_d}}} = \frac{5.0}{\frac{11.6}{\sqrt{10}}} = 1.36$$

This is an upper tail test and the p-value is the area above 1.36, in a t-distribution with df $= 9$. We see that the p-value is 0.103. Even at a 10% level, we do not find a significant difference between the two means. We do not find sufficient evidence to conclude that μ_1 is greater than μ_2.

6.241 This is a matched pairs experiment, since all 50 men get both treatments in random order and we are looking at the differences in the results. We use paired data analysis.

6.243 This is a matched pairs experiment, since the "treatment" students are each matched with a similar "control" student. We use paired data analysis.

6.245 This is a matched pairs experiment since the twins are matched and we are investigating the differences between the twins. We use paired data analysis.

6.247 We are testing $H_0 : \mu_T = \mu_N$ vs $H_a : \mu_T > \mu_N$ which we can also write as $H_0 : \mu_D = 0$ vs $H_a : \mu_d > 0$, where μ_T is mean production after drinking tea, μ_N is mean production with no tea, and μ_d is the difference

$\mu_T - \mu_N$. The original statistic is $\overline{x}_d = 293$, the standard deviation of the differences is $s_d = 242$ and the sample size is $n = 5$. The null parameter is 0 and the standardized test statistic is

$$t = \frac{\overline{x}_d - 0}{s_d/\sqrt{n_d}} = \frac{293 - 0}{242/\sqrt{5}} = 2.71$$

The p-value is found using the t-distribution with degrees of freedom 4. The p-value is 0.027 which is less than the 5% significance level. We conclude that mean production of this disease fighting molecule is higher after drinking tea.

6.249 To find a 99% confidence interval using this paired difference in means data (with degrees of freedom 49), we use $t^* = 2.68$. We have

$$\overline{x}_d \quad \pm \quad t^* \frac{s_d}{\sqrt{n_d}}$$
$$21.7 \quad \pm \quad 2.68 \cdot \frac{46.5}{\sqrt{50}}$$
$$21.7 \quad \pm \quad 17.6$$
$$4.1 \quad \text{to} \quad 39.3$$

We are 99% confident that the average decrease in testosterone level in a man who has sniffed female tears is between 4.1 and 39.3 pg/ml.

6.251 Because the data are paired, we compute a difference (Quiz pulse − Lecture pulse) for each pair. These differences are displayed in the table and dotplot below.

Student	1	2	3	4	5	6	7	8	9	10	Mean	Std. Dev.
Quiz − Lecture	+2	-1	+5	-8	+1	+20	+15	-4	+9	-12	2.7	9.93

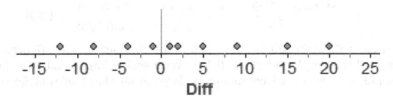

The distribution of differences appears to be relatively symmetric with no clear outliers, so we can use the t-distribution. From the *Quiz − Lecture* differences in the table above we find that the mean difference is $\overline{x}_d = 2.7$ and the standard deviation of the differences is $s_d = 9.93$. We can express the hypotheses as $H_0 : \mu_Q = \mu_L$ vs $H_a : \mu_Q > \mu_L$, where μ_Q and μ_L are the mean pulse rates during a quiz and lecture, respectively. Equivalently, we can use $H_0 : \mu_d = 0$ vs $H_a : \mu_d > 0$, where μ_d is the mean difference of quiz minus lecture pulse rates. These two ways of expressing the hypotheses are equivalent, since $\mu_d = \mu_Q - \mu_L$, and either way of expressing the hypotheses is acceptable. From the differences and summary statistics, we compute a t-statistic with

$$t = \frac{\overline{x}_d - 0}{s_d/\sqrt{n_d}} = \frac{2.7}{9.93/\sqrt{10}} = 1.52$$

To find the one-tailed p-value for this t-statistic we use the area above 1.52 in a t-distribution with $10 - 1 = 9$ degrees of freedom. This gives a p-value of 0.0814 which is not significant at a 5% level. Thus the data from these 10 students do not provide convincing evidence that mean quiz pulse rate is higher than the mean pulse rate during lecture.

6.253 (a) The data are matched by which story it is, so a matched pairs analysis is appropriate. The matched pairs analysis is particularly important here because there is a great deal of variability between the enjoyment level of the different stories.

(b) We first find the differences for all 12 stories, using the spoiler rating minus the rating for the original.

Story	1	2	3	4	5	6	7	8	9	10	11	12
With spoiler	4.7	5.1	7.9	7.0	7.1	7.2	7.1	7.2	4.8	5.2	4.6	6.7
Original	3.8	4.9	7.4	7.1	6.2	6.1	6.7	7.0	4.3	5.0	4.1	6.1
Difference	0.9	0.2	0.5	-0.1	0.9	1.1	0.4	0.2	0.5	0.2	0.5	0.6

We compute the summary statistics for the differences: $\bar{x}_d = 0.492$ with $s_d = 0.348$ and $n_d = 12$. Using a t-distribution with df = 11 for a 95% confidence level, we have $t^* = 2.20$. We find the 95% confidence interval:

$$\bar{x}_d \pm t^* \cdot \frac{s_d}{\sqrt{n_d}}$$

$$0.492 \pm 2.20 \cdot \frac{0.348}{\sqrt{12}}$$

$$0.492 \pm 0.221$$

$$0.271 \text{ to } 0.713$$

We are 95% sure that mean enjoyment rating of a version of a story *with* a spoiler is between 0.271 to 0.713 higher than the mean enjoyment rating of the original story without it.

6.255 Since each baby received both the speech and singing treatment this is a matched pairs design, and since we are not testing specific direction this is a two-tailed test, $H_0 : \mu_d = 0$ vs $H_a : \mu_d \neq 0$, where μ_d is the mean difference (speaking minus singing). The matched pairs test statistic is

$$t = \frac{\text{Statistic} - \text{Null value}}{SE} = \frac{\bar{x}_d - 0}{s_d/\sqrt{n_d}} = \frac{10.39 - 0}{55.37/\sqrt{48}} = 1.300$$

We find the proportion in the right tail beyond 1.300 of a t-distribution with 47 degrees of freedom to be 0.100. This is a two-tail test, so the p-value is 0.200. With such a large p-value we fail to reject the null hypothesis and conclude that we don't have enough evidence to say that babies prefer speaking over singing or vice-versa.

6.257 (a) This is a difference in means test with separate samples. We first compute the summary statistics and see that, for the first quiz, $n_1 = 10$ with $\bar{x}_1 = 78.6$ with $s_1 = 13.4$ and, for the second quiz, $n_2 = 10$ with $\bar{x} = 84.0$ with $s_2 = 8.1$. To conduct the test for the difference in means, the hypotheses are:

$$H_0 : \quad \mu_1 = \mu_2$$
$$H_a : \quad \mu_1 < \mu_2$$

where μ_1 is the average grade for all students on the first quiz and μ_2 is the average grade for all students on the second quiz. The test statistic is

$$\frac{\bar{x}_1 - \bar{x}_2}{\sqrt{\frac{s_1^2}{n_1} + \frac{s_2^2}{n_2}}} = \frac{78.6 - 84.0}{\sqrt{\frac{13.4^2}{10} + \frac{8.1^2}{10}}} = -1.09$$

Using a t-distribution with df=9, we see that the p-value for this lower tail test is 0.152. This is not significant, even at a 10% level. We do not reject H_0 and do not find convincing evidence that the grades on the second quiz are higher.

(b) This is a paired difference in means test. We begin by finding the differences: first quiz − second quiz.

$$-6 \quad -1 \quad -16 \quad -2 \quad 0 \quad 3 \quad -12 \quad -2 \quad -5 \quad -13$$

The summary statistics for these 10 differences are $n_d = 10$, $\bar{x}_d = -5.4$ with $s_d = 6.29$. The hypotheses are the same as for part (a) and the t-statistic is

$$t = \frac{\bar{x}_d}{s_d/\sqrt{n_d}} = \frac{-5.4}{6.29/\sqrt{10}} = -2.71.$$

Using a t-distribution with df $= 9$, we see that the p-value for this lower-tail test is 0.012. This is significant at a 5% level and almost at even a 1% level. We reject H_0 and find evidence that mean grades on the second quiz are higher.

(c) The spread of the grades is very large on both quizzes, so the high variability makes it hard to find a difference in means with the separate samples. Once we know that the data are paired, it eliminates the variability between people. In this case, it is much better to collect the data using a matched pairs design.

Unit C: Essential Synthesis Solutions

C.1 We are estimating mean amount spent in the store, so we use a confidence interval for a mean.

C.3 We are testing for a difference in the proportion classified as insulin-resistant between high sugar and normal diets, so we use a hypothesis test for a difference in proportions.

C.5 We are estimating the difference in the mean financial aid package between two groups of students, so we use a confidence interval for a difference in means.

C.7 We are testing whether the proportion of left-handers is different from 12%, so we use a hypothesis test for a proportion.

C.9 The z test statistic can be interpreted as a z-score so this test statistic is more than 5 standard deviations out in the tail. This will have very little area beyond it (so the p-value is very small) and will provide strong evidence for the alternative hypothesis (so we reject H_0).

  (a) Small

  (b) Reject H_0

C.11 The z test statistic can be interpreted as a z-score so this test statistic is less than 1 standard deviation from the mean. This will have quite a large area beyond it (so the p-value is relatively large) and will not provide much evidence for the alternative hypothesis (so we do not reject H_0).

  (a) Large

  (b) Do not reject H_0

C.13 For relatively large sample size, the t-distribution is very similar to the standard normal distribution, so the t test statistic is similar to a z-score. This test statistic is approximately 7 standard deviations out in the tail! This will have very little area beyond it (so the p-value is very small) and will provide strong evidence for the alternative hypothesis (so we reject H_0).

  (a) Small

  (b) Reject H_0

C.15 It is best to begin by adding the totals to the two-way table.

	A	B	C	Total
Yes	21	15	15	51
No	39	50	17	106
Total	60	65	32	157

If we let p be the proportion of bills by Server B, we are testing $H_0 : p = 1/3$ vs $H_a : p > 1/3$. We have $\hat{p} = 65/157 = 0.414$. The sample size is large enough to use the normal distribution. We have

$$z = \frac{\hat{p} - p_0}{\sqrt{\frac{p_0(1-p_0)}{n}}} = \frac{0.414 - 1/3}{\sqrt{\frac{1/3(2/3)}{157}}} = 2.14$$

This is a upper-tail test so the p-value is the area above 2.14 in a standard normal distribution, so we see that the p-value is 0.0162. At a 5% significance level, we reject H_0 and conclude that there is evidence that Server B is responsible for more than 1/3 of the bills at this restaurant. The results, however, are not strong enough to be significant at a 1% level.

C.17 This is a test for a difference in proportions. We are testing $H_0 : p_B = p_C$ vs $H_a : p_B \neq p_C$, where p_B and p_C are the proportions of bills paid with cash for Server B and Server C, respectively. We see from the table that $\hat{p}_B = 50/65 = 0.769$ and $\hat{p} = 17/32 = 0.531$. The sample sizes are large enough to use the normal distribution, and we compute the pooled proportion as $\hat{p} = 67/97 = 0.691$. We have

$$z = \frac{(\hat{p}_B - \hat{p}_C) - 0}{\sqrt{\hat{p}(1 - \hat{p})(\frac{1}{n_B} + \frac{1}{n_C})}} = \frac{0.769 - 0.531}{\sqrt{0.691(0.309)(\frac{1}{65} + \frac{1}{32})}} = 2.39$$

This is a two-tail test, so the p-value is twice the area above 2.39 in a standard normal distribution. We see that

$$\text{p-value} = 2(0.0084) = 0.0168.$$

At a 5% significance level, we reject H_0 and conclude that there is evidence that the proportion paying with cash is not the same between Server B and Server C. Server B appears to have a greater proportion of customers paying with cash. The results, however, are not strong enough to be significant at a 1% level.

C.19 In the sample, the bill is larger when paying with a credit or debit card ($\overline{x}_Y = 29.4 > \overline{x}_N = 19.5$) and there is more variability with a card ($s_Y = 14.5 < s_N = 9.4$). To determine if there is evidence of a difference in the mean bill depending on the method of payment, we do a test for a difference in means. We are testing $H_0 : \mu_Y = \mu_N$ vs $H_a : \mu_Y \neq \mu_N$, where μ_Y represents the mean bill amount when paying with a credit or debit card and μ_N represents the mean bill amount when paying with cash. The sample sizes are large enough to use the t-distribution, even if the underlying bills are not normally distributed. We have

$$t = \frac{(\overline{x}_Y - \overline{x}_N) - 0}{\sqrt{\frac{s_Y^2}{n_Y} + \frac{s_N^2}{n_N}}} = \frac{29.4 - 19.5}{\sqrt{\frac{14.5^2}{51} + \frac{9.4^2}{106}}} = 4.45$$

This is a two-tail test, so the p-value is twice the area above 4.45 in a t-distribution with $51 - 1 = 50$ df. This area is essentially zero, so we have p-value ≈ 0. At any reasonable significance level, we find strong evidence to reject H_0. There is strong evidence that the mean bill amounts are not the same between the two payment methods. Customers with higher bills are more likely to pay with a credit/debit card.

C.21 None of the data appears to be severely skewed or with significant outliers, so it is fine to use the t-distribution. We use technology and the data in **MentalMuscle** to obtain the means and standard deviations for the times in each group as summarized in the table below.

Variable	Action	Fatigue	N	Mean	StDev
Time	Mental	Pre	8	7.34	1.22
Time	Mental	Post	8	6.10	0.95
Time	Actual	Pre	8	7.16	0.70
Time	Actual	Post	8	8.04	1.07

(a) To compare the mean mental and actual times before fatigue we do a two-sample t-test for difference in means. Letting μ_M represent the mean time for someone mentally imaging the actions before any muscle fatigue and μ_A represent the mean time for someone actually performing the actions before any muscle fatigue, our hypotheses are:

$$H_0 : \quad \mu_M = \mu_A$$
$$H_a : \quad \mu_M \neq \mu_A$$

Using the pre-fatigue data values in each case, the relevant summary statistics as $\bar{x}_M = 7.34$ with $s_M = 1.22$ and $n_M = 8$ for the mental pre-fatigue group and $\bar{x}_A = 7.16$ with $s_A = 0.70$ and $n_A = 8$ for the actual pre-fatigue group. The test statistic is

$$t = \frac{\bar{x}_M - \bar{x}_A}{\sqrt{\frac{s_M^2}{n_M} + \frac{s_A^2}{n_A}}} = \frac{7.34 - 7.16}{\sqrt{\frac{1.22^2}{8} + \frac{0.70^2}{8}}} = 0.36$$

This is a two-tail test, so the p-value is twice the area above 0.36 in a t-distribution with $df = 7$. We see that the p-value is $2(0.365) = 0.73$. This is a very large p-value, so we do not reject H_0. There is no convincing evidence at all of a difference between the two groups before muscle fatigue.

(b) For each action, the same 8 people perform the movements twice, once before muscle fatigue and once after. To compare the pre-fatigue and post-fatigue means for those actually doing the movements we use a paired difference in means test. We compute the differences, $D = PostFatigue - PreFatigue$, using the data for the people doing the actual movements and see that the 8 differences are:

$$2.5, \quad 0, \quad 0.7, \quad 0.5, \quad 0.2, \quad 0.6, \quad -0.3, \quad 2.8$$

The summary statistics for the differences are $\bar{x}_D = 0.88$ with $s_D = 1.15$ and $n_D = 8$. The hypotheses are

$$H_0: \quad \mu_D = 0$$
$$H_a: \quad \mu_D > 0$$

Notice that if we had subtracted the other direction ($PreFatigue - PostFatigue$), the alternative would be in the other direction and all the differences would have the opposite sign, but the results would be identical. The test statistic is

$$t = \frac{\bar{x}_D}{s_D/\sqrt{n_D}} = \frac{0.88}{1.15/\sqrt{8}} = 2.16$$

This is an upper-tail test, so the p-value is the area above 2.16 in a t-distribution with $df = 7$. Using technology we see that the p-value is 0.0338. At a 5% significance level, we reject H_0 and conclude that people are slower, on average, at performing physical motions when they have muscle fatigue. This is not surprising; we slow down when we are tired!

(c) As in part (b), the same 8 people perform the movements twice, once before muscle fatigue and once after. To compare the pre-fatigue and post-fatigue means for those doing mental movements we use a paired difference in means test. We compute the differences, $D = PostFatigue - PreFatigue$, using the data for the people doing the mental imaging and see that the 8 differences are:

$$1.5, \quad -3.9, \quad -1.5, \quad -1.1, \quad -1.1, \quad -0.2, \quad -1.5, \quad -2.1$$

The summary statistics for the differences are $\bar{x}_D = -1.24$ with $s_D = 1.54$ and $n_D = 8$. The hypotheses are;

$$H_0: \quad \mu_D = 0$$
$$H_a: \quad \mu_D < 0$$

Notice that if we had subtracted the other direction ($PreFatigue - PostFatigue$), the alternative would be in the other direction and all the differences would have the opposite sign, but the results would be identical. The test statistic is

$$t = \frac{\bar{x}_D}{s_D/\sqrt{n_D}} = \frac{-1.24}{1.54/\sqrt{8}} = -2.28$$

This is a lower-tail test, so the p-value is the area below -2.28 in a t-distribution with $df = 7$. Using technology we see that the p-value is 0.0283. At a 5% significance level, we reject H_0 and conclude that people are faster, on average, at mentally imaging physical motions when they have muscle fatigue. This is the new finding from the study: when people have muscle fatigue, they speed up their mental imaging – presumable because they just want to get done! It is likely that this makes the mental imagery less effective.

(d) There are two separate groups (*Mental* vs *Actual*) so we do a difference in means test with two groups. Letting μ_M represent the mean time for someone mentally imaging the actions after muscle fatigue and μ_A represent the mean time for someone actually performing the actions after muscle fatigue, our hypotheses are:

$$H_0: \quad \mu_M = \mu_A$$
$$H_a: \quad \mu_M \neq \mu_A$$

Using the post-fatigue data values in each case, the relevant summary statistics are $\overline{x}_M = 6.10$ with $s_M = 0.95$ and $n_M = 8$ for the mental post-fatigue group and $\overline{x}_A = 8.04$ with $s_A = 1.07$ and $n_A = 8$ for the actual post-fatigue group. The test statistic is

$$t = \frac{\overline{x}_M - \overline{x}_A}{\sqrt{\frac{s_M^2}{n_M} + \frac{s_A^2}{n_A}}} = \frac{6.10 - 8.04}{\sqrt{\frac{0.95^2}{8} + \frac{1.07^2}{8}}} = -3.83.$$

This is a two-tail test, so the p-value is twice the area below -3.83 in a t-distribution with $df = 7$. We see that the p-value is $2(0.0032) = 0.0064$. This is a very small p-value, so we reject H_0. There is strong evidence of a difference in the mean times between the two groups after muscle fatigue.

(e) Before muscle fatigue, the group mentally imaging doing the actions was remarkably similar in time to those actually doing the actions. The mental imaging was quite accurate. However, muscle fatigue caused those actually doing the motions to slow down while it caused those mentally imaging the motions to speed up. Taken together, there was a significant difference between the two groups after experiencing muscle fatigue so that the mental imaging of the motions was not as accurate at matching the actual motions.

C.23 If μ is the mean number of free throws attempted in games by the Warriors, we test $H_0 : \mu = 25.0$ vs $H_a : \mu \neq 25.0$. For the sample of 82 games in 2015-2016 in **GSWarriors** we find the mean number of free throw attempts by the Warriors is $\overline{x} = 21.83$ with a standard deviation of 7.73. The t-statistic is

$$t = \frac{\overline{x} - \mu_0}{s/\sqrt{n}} = \frac{21.83 - 25.0}{7.73/\sqrt{82}} = -3.71$$

We find the p-value by doubling the area beyond -3.71 in a t-distribution with $82 - 1 = 81$ degrees of freedom, p-value $= 2 \cdot 0.0019 = 0.0038$. This is a very small p-value, so we have strong evidence that the Golden State Warriors average fewer free throw attempts per game than is typical for NBA teams.

C.25 This question is asking for a test to compare two proportions, p_H and p_A, the proportion of free throws the Warriors make in home and away games, respectively. The question also suggests a particular direction so the hypotheses are $H_0 : p_H = p_A$ vs $H_a : p_H > p_A$. Based on the sample results we get the proportions below for each location and the combined data.

$$\hat{p}_H = \frac{700}{894} = 0.783 \qquad \hat{p}_A = \frac{666}{896} = 0.743 \qquad \hat{p} = \frac{1366}{1790} = 0.763$$

The standardized test statistic is

$$z = \frac{0.783 - 0.743}{\sqrt{\frac{0.763(1-0.763)}{894} + \frac{0.763(1-0.763)}{896}}} = \frac{0.040}{0.020} = 2.0$$

Since the sample sizes are large we find a p-value using the area in a $N(0,1)$ distribution that lies beyond $z = 2.0$. This gives p-value= 0.023 which is small (at a 5% level), so we have sufficient evidence to conclude that the Warriors make a higher proportion of their free throws at home than they do on the road.

C.27 (a) The hypotheses are

$$H_0: \quad \mu = 72$$
$$H_a: \quad \mu \neq 72$$

where μ represents the average heart rate of all patients admitted to this ICU. Using technology, we see that the average heart rate for the sample of patients is $\bar{x} = 98.92$ and we see that the p-value for the test is essentially zero (with a t-statistic of 14.2). There is very strong evidence that the average heart rate of ICU patients is not 72.

(b) Using technology, we see that 40 of the 200 patients died, so $\hat{p} = 40/200 = 0.2$. Using technology, we see that a 95% confidence interval for the proportion who die is $(0.147, 0.262)$. We are 95% confident that between 14.7% and 26.2% of the ICU patients die at this hospital.

(c) We see that there were 124 females (62%) and 76 males (38%), so more females were admitted to the ICU in this sample. To test whether genders are equally split, we do a one-sample proportion test. It doesn't matter whether we test for the proportion of males or the proportion of females, since if we know one, we can compute the other, and in either case we are testing whether the proportion is significantly different from 0.5. Using p to denote the proportion of ICU patients that are female, we test

$$H_0: \quad p = 0.5$$
$$H_a: \quad p \neq 0.5$$

Using technology, we see that the p-value for this test is 0.001. This provides very strong evidence that patients are not evenly split between the genders. There are significantly more females than males admitted to this ICU unit.

(d) This is a difference in means test. If we let μ_M represent the mean age of male ICU patients and μ_F represent the mean age of female ICU patients, the hypotheses are

$$H_0: \quad \mu_M = \mu_F$$
$$H_a: \quad \mu_M \neq \mu_F$$

Using technology, we see that the p-value is 0.184. We do not reject H_0 and do not find convincing evidence that mean age differs between males and females.

(e) This is a difference in proportions test. If we let p_M represent the proportion of males who die and p_F represent the proportion of females who die, the hypotheses are

$$H_0: \quad p_M = p_F$$
$$H_a: \quad p_M \neq p_F$$

Using technology, we see that the p-value is 0.772. We do not reject H_0 and do not find convincing evidence that the proportion who die differs between males and females.

Unit C: Review Exercise Solutions

C.29 Using technology on a calculator or computer, we see that

(a) The area below $z = -2.10$ is 0.018.

(b) The area above $z = 1.25$ is 0.106.

C.31 Using technology on a calculator or computer, we see that the endpoint z is

(a) $z = 0.253$

(b) $z = -2.054$

C.33 We use a t-distribution with df $= 24$. Using technology, we see that the values with 5% beyond them in each tail are ± 1.711.

C.35 We use a t-distribution with df $= 9$. Using technology, we see that the area above 2.75 is 0.011.

C.37 (a) To find a confidence interval using the normal distribution, we use

$$\text{Sample statistic} \pm z^* \cdot SE.$$

We are finding a confidence interval for a mean, so the statistic from the original sample is $\bar{x} = 12.79$. For a 95% confidence interval, we use $z^* = 1.960$ and we have $SE = 0.30$. Putting this information together, we have

$$
\begin{array}{ccc}
\bar{x} & \pm & z^* \cdot SE \\
12.79 & \pm & 1.960 \cdot (0.30) \\
12.79 & \pm & 0.59 \\
12.20 & \text{to} & 13.38
\end{array}
$$

We are 95% confident that the average number of grams of fiber consumed per day is 12.20 grams and 13.38 grams.

(b) The relevant hypotheses are $H_0 : \mu = 12$ vs $H_a : \mu > 12$, where μ is the mean number of grams of fiber consumed per day by all people in the population. The statistic of interest is $\bar{x} = 12.79$. The standard error of this statistic is given as $SE = 0.30$ and the null hypothesis is that the population mean is 12. We compute the standardized test statistic with

$$z = \frac{\text{Sample Statistic} - \text{Null Parameter}}{SE} = \frac{12.79 - 12}{0.30} = 2.63$$

Using technology, the area under a $N(0, 1)$ curve beyond $z = 2.63$ is 0.004. This small p-value provides strong evidence that the mean number of grams of fiber consumed per day is greater than 12.

C.39 The sample size is definitely large enough to use the normal distribution. For a confidence interval using the normal distribution, we use

$$\text{Sample statistic} \pm z^* \cdot SE.$$

The relevant sample statistic for a confidence interval for a proportion is $\hat{p} = 0.83$. For a 95% confidence interval, we have $z^* = 1.96$, and the standard error is $SE = \sqrt{\hat{p}(1-\hat{p})/n}$. The confidence interval is

$$\hat{p} \quad \pm \quad z^*\sqrt{\frac{\hat{p}(1-\hat{p})}{n}}$$

$$0.83 \quad \pm \quad 1.96 \cdot \sqrt{\frac{0.83(0.17)}{1000}}$$

$$0.83 \quad \pm \quad 0.023$$

$$0.807 \quad \text{to} \quad 0.853$$

We are 95% confident that the proportion of adults who believe that children spend too much time on electronic devices is between 0.807 and 0.853. The margin of error is 0.023. Since the lowest plausible value for p in the confidence interval is 0.807, it is *not* plausible that the proportion is less than 80%. Since 0.85 is within the plausible range in the confidence interval, it is plausible that the proportion is greater than 85%.

C.41   (a) We see that $n = 157$, $\bar{x} = 3.849$, and $s = 2.421$.

(b) We have

$$SE = \frac{s}{\sqrt{n}} = \frac{2.421}{\sqrt{157}} = 0.193$$

This is the same as the value given in the computer output.

(c) For a 95% confidence interval with degrees of freedom 156, we use $t^* = 1.98$. The confidence interval is

$$\bar{x} \quad \pm \quad t^* \cdot \frac{s}{\sqrt{n}}$$

$$3.849 \quad \pm \quad 1.98 \cdot \frac{2.421}{\sqrt{157}}$$

$$3.849 \quad \pm \quad 0.383$$

$$3.466 \quad \text{to} \quad 4.232$$

A 95% confidence interval is \$3.466 to \$4.232.

(d) Up to two decimal places, the confidence interval we found is the same as the one given in the computer output. The small differences are probably due to round-off error in estimating the t^* value.

(e) We are 95% confident that the average tip given at this restaurant is between \$3.47 and \$4.23.

C.43   (a) The cookies were bought from locations all over the country to try to avoid sampling bias.

(b) Let μ be the mean number of chips per bag. We are testing $H_0 : \mu = 1000$ vs $H_a : \mu > 1000$. The test statistic is

$$t = \frac{1261.6 - 1000}{117.6/\sqrt{42}} = 14.4$$

We use a t-distribution with 41 degrees of freedom. The area to the left of 14.4 is negligible, and p-value ≈ 0. We conclude, with very strong evidence, that the average number of chips per bag of Chips Ahoy! cookies is greater than 1000.

(c) No! The test in part (b) gives convincing evidence that the *average* number of chips per bag is greater than 1000. However, this does not necessarily imply that every individual bag has more than 1000 chips.

C.45 (a) Using I for the internet users and N for the non-internet users, we see that $\hat{p}_I = 807/1754 = 0.46$ and $\hat{p}_N = 130/483 = 0.27$. In the sample, the internet users are more trusting.

(b) A hypothesis test is used to determine whether we can generalize the results from the sample to the population. Since we are looking for a difference, this is a two-tail test. The hypotheses are

$$H_0: \quad p_I = p_N$$
$$H_a: \quad p_I \neq p_N$$

In addition to the two sample proportions computed in part (a), we compute the pooled proportion. A total of $807 + 130$ people in the full sample of $1754 + 483$ people agreed with the statement, so

$$\text{Pooled proportion} = \hat{p} = \frac{807 + 130}{1754 + 483} = \frac{937}{2237} = 0.419.$$

The standardized test statistic is

$$z = \frac{\hat{p}_I - \hat{p}_N}{\sqrt{\frac{\hat{p}(1-\hat{p})}{n_I} + \frac{\hat{p}(1-\hat{p})}{n_N}}} = \frac{0.46 - 0.27}{\sqrt{\frac{0.419(0.581)}{1754} + \frac{0.419(0.581)}{483}}} = 7.49.$$

This is a two-tail test, so the p-value is two times the area above 7.49 in a standard normal distribution. However, the area above 7.49 in a standard normal is essentially zero (more than seven standard deviations above the mean!). The p-value is essentially zero. The p-value is extremely small, and gives very strong evidence that internet users are more trusting than non-internet users.

(c) No, we cannot conclude that internet uses causes people to be more trusting. The data come from an observational study rather than a randomized experiment. There are many possible confounding factors.

(d) Yes. Level of formal education is a confounding factor if education level affects whether a person uses the internet (it does – more education is associated with more internet use) and also if education level affects how trusting someone is (it does – more education is associated with being more trusting). Remember that a confounding factor is a factor that influences both of the variables of interest. (In fact, even after controlling for education level and several other confounding variables, the data still show that internet users are more trusting than non-users.)

C.47 For the difference in mean home price between California and New York, we use

$$(\bar{x}_{ca} - \bar{x}_{ny}) \pm t^* \sqrt{\frac{s_{ca}^2}{n_{ca}} + \frac{s_{ny}^2}{n_{ny}}}$$

Both samples have size $n_{ca} = n_{ny} = 30$ so we use 29 degrees of freedom to find $t^* = 1.699$ for 90% confidence. The confidence interval is

$$
\begin{aligned}
(715.1 - 565.6) \quad &\pm \quad 1.699\sqrt{\frac{1112.3^2}{30} + \frac{697.6^2}{30}} \\
149.5 \quad &\pm \quad 407.3 \\
-257.8 \quad &\text{to} \quad 556.8
\end{aligned}
$$

We are 90% sure that the mean home price in California is somewhere between $257.8 thousand dollars less than in New York to as much as $556.8 thousand more.

C.49 For the difference in mean home price between New York and New Jersey, we use

$$(\bar{x}_{ny} - \bar{x}_{nj}) \pm t^* \sqrt{\frac{s_{ny}^2}{n_{ny}} + \frac{s_{nj}^2}{n_{nj}}}$$

Both samples have size $n_{ny} = n_{nj} = 30$ so we use 29 degrees of freedom to find $t^* = 2.045$ for 95% confidence. The confidence interval is

$$
\begin{aligned}
(565.6 - 388.5) \;\; &\pm \;\; 2.045 \sqrt{\frac{697.5^2}{30} + \frac{224.7^2}{30}} \\
177.1 \;\; &\pm \;\; 273.6 \\
-96.5 \;\; &\text{to} \;\; 450.7
\end{aligned}
$$

We are 95% sure that the mean home price in New York is somewhere between \$96.5 thousand dollars less than in New Jersey to as much as \$450.7 thousand more.

C.51 Let μ_C and μ_I represent mean weight loss after six months for women on a continuous or intermittent calorie restricted diet, respectively. The hypotheses are: The hypotheses are:

$$
\begin{aligned}
H_0 : \;\; & \mu_C = \mu_I \\
H_a : \;\; & \mu_C \neq \mu_I
\end{aligned}
$$

The relevant statistic for this test is $\bar{x}_C - \bar{x}_I$, and the relevant null parameter is zero, since from the null hypothesis we have $\mu_C - \mu_I = 0$. The t-test statistic is:

$$t = \frac{\text{Sample statistic} - \text{Null parameter}}{SE} = \frac{(\bar{x}_C - \bar{x}_I) - 0}{\sqrt{\frac{s_C^2}{n_C} + \frac{s_I^2}{n_I}}} = \frac{14.1 - 12.2}{\sqrt{\frac{13.2^2}{54} + \frac{10.6^2}{53}}} = 0.82$$

This is a two-tail test, so the p-value is two times the area above 0.82 in a t-distribution with df $= 52$. We see that the p-value is $2(0.208) = 0.416$. This p-value is not small at all so we do not reject H_0. We do not see convincing evidence of a difference in effectiveness of the two weight loss methods.

C.53 We are testing for a difference in two means from a matched pairs experiment. We can use as our null hypothesis either $H_0 : \mu_{on} = \mu_{off}$ or, equivalently, $H_0 : \mu_d = 0$. The two are equivalent since, using d to represent the differences, we have $\mu_d = \mu_{on} - \mu_{off}$. Using the differences, we have

$$
\begin{aligned}
H_0 : \;\; & \mu_d = 0 \\
H_a : \;\; & \mu_d > 0
\end{aligned}
$$

Notice that this is a one-tail test since we are specifically testing to see if the metabolism is higher for the "on" condition. The t-test statistic is:

$$t = \frac{\bar{x}_d - 0}{s_d / \sqrt{n_d}} = \frac{2.4}{6.3 / \sqrt{47}} = 2.61$$

Using the t-distribution with degrees of freedom 46, we find a p-value of 0.006. This provides enough evidence to reject H_0 and conclude that mean brain metabolism is significantly higher when a cell phone is turned on and pressed to the ear.

C.55 (a) The best estimate is the sample proportion, 0.17. Using $z^* = 2.576$ for 99% confidence, the margin of error is

$$ME = z^* \cdot \sqrt{\frac{\hat{p}(1-\hat{p})}{n}} = 2.576 \cdot \sqrt{\frac{0.17(0.83)}{1016}} = 0.03.$$

The margin of error is $\pm 3\%$.

(b) We use $z^* = 2.576$ for 99% confidence, and the sample proportion $\hat{p}$ for our estimated proportion $\tilde{p} = 0.17$. If we want the margin of error to be within $\pm 1\%$, we need a sample size of:

$$n = \left(\frac{2.576}{0.01}\right)^2 (0.17)(1 - 0.17) = 9363.1$$

We round up to require at least 9,364 US adults in the sample.

C.57 These are large sample sizes so the normal distribution is appropriate. For 95% confidence $z^* = 1.96$. The sample proportions are $\hat{p}_H = 164/8506 = 0.0193$ and $\hat{p}_P = 122/8102 = 0.0151$ in the HRT group and the placebo group, respectively. The confidence interval is

$$(0.0193 - 0.0151) \pm 1.96 \cdot \sqrt{\frac{0.0193(1 - 0.0193)}{8506} + \frac{0.0151(1 - 0.0151)}{8102}} = 0.0042 \pm 0.0039 = (0.0003, 0.0081).$$

We are 95% sure that the proportion of women who get cardiovascular disease is between 0.0003 and 0.0081 higher among women who get hormone replacement therapy rather than a placebo.

C.59 These are large sample sizes so the normal distribution is appropriate. For 95% confidence $z^* = 1.96$. The sample proportions are $\hat{p}_H = 502/8506 = 0.0590$ and $\hat{p}_P = 458/8102 = 0.0565$ in the HRT group and the placebo group, respectively. The confidence interval is

$$(0.0590 - 0.0565) \pm 1.96 \cdot \sqrt{\frac{0.0590(1 - 0.0590)}{8506} + \frac{0.0565(1 - 0.0565)}{8102}} = 0.0025 \pm 0.0071 = (-0.0046, 0.0096).$$

We are 95% sure that the proportion of women who get any form of cancer is between 0.0046 lower and 0.0096 higher among women who get hormone replacement therapy rather than a placebo.

C.61 (a) The sample for New York has only $0.267 \times 30 = 8$ homes with more then 3 bedrooms, so the normal distribution may not be appropriate.

(b) Because the normal distribution may not apply, we return to the methods of Chapter 4 and perform a randomization test for the hypotheses $H_0 : p_{NY} = p_{NJ}$ vs $H_a : p_{NY} \neq p_{NJ}$. We use resampling, taking samples with replacement from the pooled data, because the randomness in this problem comes from sampling, not any kind of allocation. Using *StatKey* or other technology we construct a randomization distribution of many differences in proportions for simulated samples of size 30, taken from a "population" with the same proportion of three bedroom homes as we see in the combined sample.

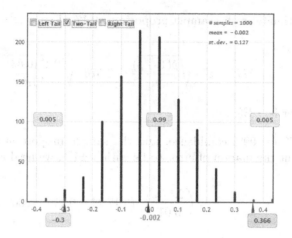

In this randomization distribution we see that 5 of the 1000 samples give differences as extreme as the difference of $0.633 - 0.267 = 0.366$ that occurred in the original sample. Because this is a two-sided test, we double the 0.005 to get p-value = 0.01. We reject the null hypothesis, and conclude that the proportion of homes with more then 3 bedrooms is different between New York and New Jersey.

C.63 Let μ be the mean volume of juice per bottle, in fl oz. We are testing $H_0 : \mu = 12.0$ vs $H_a : \mu \neq 12.0$. The test statistic is

$$t = \frac{11.92 - 12.0}{0.26/\sqrt{30}} = -1.69.$$

We use a t-distribution with 29 degrees of freedom. The area to the left of -1.69 is 0.051, so the p-value is $2(0.051) = 0.102$. We do not reject H_0 at the 1% level. There is not sufficient evidence that the average amount of juice per bottle differs from 12 fl oz, so Susan need not recalibrate the machine.

C.65 We estimate the difference in population means using the difference in sample means $\overline{x}_L - \overline{x}_D$, where $\overline{x}_L$ represents the mean weight gain of the mice in light and $\overline{x}_D$ represents the mean weight gain of mice in darkness. For a 99% confidence interval with degrees of freedom equal to 7, we use $t^* = 3.50$. We have:

$$
\begin{aligned}
\text{Sample statistic} \quad &\pm \quad t^* \cdot SE \\
(\overline{x}_L - \overline{x}_D) \quad &\pm \quad t^* \sqrt{\frac{s_L^2}{n_L} + \frac{s_D^2}{n_D}} \\
(9.4 - 5.9) \quad &\pm \quad 3.50\sqrt{\frac{3.2^2}{19} + \frac{1.9^2}{8}} \\
3.5 \quad &\pm \quad 3.48 \\
0.02 \quad &\text{to} \quad 6.98
\end{aligned}
$$

We are 99% confident that mice with light at night will gain, on average, between 0.02 grams and 6.98 grams more than mice in darkness.

C.67 Depending on technology, we may need to add a new column to the **NFLScores2011** dataset to compute the difference in home and away scores with a formula like $Diff = HomeScore - AwayScore$. After doing so, we can find the summary statistics for the differences as in the output below.

```
Variable   N   Mean   StDev  Minimum  Maximum
Diff      256  3.266  15.203  -35.000  55.000
```

Within this sample of $n_d = 256$ games, the mean difference is $\bar{x}_d = 3.27$ points in favor of the home team with a standard deviation of $s_d = 15.2$. For 90% confidence with a t-distribution and 255 df we have $t^* = 1.65$. To compute the confidence interval based on the paired data we have

$$\bar{x}_d \quad \pm \quad t^* \frac{s_d}{\sqrt{n_d}}$$
$$3.27 \quad \pm \quad 1.65 \cdot \frac{15.2}{\sqrt{256}}$$
$$3.27 \quad \pm \quad 1.57$$
$$1.70 \quad \text{to} \quad 4.84$$

Based on these results we are 90% sure that the mean home field advantage for NFL games is between 1.70 and 4.84 points. Football fans often use the adage that the home field is worth about a field goal which counts for 3 points. That would certainly be consistent with this interval.

C.69 First, we compute the proportion of students choosing the Olympic gold medal within each of the gender samples.

$$\hat{p}_m = \frac{109}{193} = 0.565 \quad \text{and} \quad \hat{p}_f = \frac{73}{169} = 0.432$$

The estimated difference in proportions is $\hat{p}_m - \hat{p}_f = 0.565 - 0.432 = 0.133$. The sample sizes are both quite large, well more than 10 students choosing each type of award within each gender, so we model the differences in proportions with a normal distribution. For 90% confidence the standard normal endpoint is $z^* = 1.645$. This gives

$$(\hat{p}_m - \hat{p}_f) \quad \pm \quad z^* \cdot \sqrt{\frac{\hat{p}_m(1 - \hat{p}_m)}{n_m} + \frac{\hat{p}_f(1 - \hat{p}_f)}{n_f}}$$
$$(0.565 - 0.432) \quad \pm \quad 1.645\sqrt{\frac{0.565(1 - 0.565)}{193} + \frac{0.432(1 - 0.432)}{169}}$$
$$0.133 \quad \pm \quad 0.086$$
$$0.047 \quad \text{to} \quad 0.219$$

We are 90% sure that the proportion of male statistics students who prefer the Olympic gold medal is between 0.047 and 0.219 more than the proportion of female statistics students who make that choice.

C.71 (a) Let μ_1 represent the mean midterm grade for students who attend class on a Friday before break and μ_2 be the mean grade for students who miss that class. The instructor's suspicion is in a particular direction so we use a one-tail alternative: $H_0 : \mu_1 = \mu_2$ vs $H_a : \mu_1 > \mu_2$.

(b) The value of the t-statistic is $t = \dfrac{80.9 - 68.2}{\sqrt{\dfrac{11.07^2}{15} + \dfrac{9.26^2}{9}}} = \dfrac{12.7}{4.21} = 3.02.$

We find a p-value using a t-distribution with $9 - 1 = 8$ degrees of freedom. The area in the upper tail of this distribution beyond 3.02 gives a p-value of 0.008. This is a very small p-value so we find strong evidence to support the claim that the mean midterm grade is higher for students who attend class on the Friday before break than for those who skip.

(c) Even though the test shows strong evidence that the mean midterm grade is lower for those that skip class, we can't conclude that missing class causes this to happen. In fact, the midterm grades were determined *before* students even came to class (or chose to miss). This was not an experiment (we didn't randomly assign students to attend or miss class!) so we can't draw a cause/effect conclusion.

(d) It was a good idea to exclude the student who was never attending class. The instructor would probably like to draw a conclusion about the population of students who are regular members of the class. Also, including the extremely low midterm grade from this student would bring into question the appropriateness of using the t-distribution for this test since the sample sizes are somewhat small.

C.73 In this case, we have $p = 0.651$ and $n = 50$. The sample proportions will be centered at the population proportion of $p = 0.651$ so will have a mean of 0.651. The standard deviation of the sample proportions is the standard error, which is

$$SE = \sqrt{\frac{p(1 - p)}{n}} = \sqrt{\frac{0.651(1 - 0.651)}{50}} = 0.067$$

C.75 (a) The mean of the distribution is 36.78 years old. The standard deviation of the distribution of sample means is the standard error:

$$SE = \frac{\sigma}{\sqrt{n}} = \frac{22.58}{\sqrt{10}} = 7.14 \text{ years}$$

(b) The mean of the distribution is 36.78 years old. The standard deviation of the distribution of sample means is the standard error:
$$SE = \frac{\sigma}{\sqrt{n}} = \frac{22.58}{\sqrt{100}} = 2.258 \text{ years}$$

(c) The mean of the distribution is 36.78 years old. The standard deviation of the distribution of sample means is the standard error:
$$SE = \frac{\sigma}{\sqrt{n}} = \frac{22.58}{\sqrt{1000}} = 0.714 \text{ years}$$

Notice that as the sample size goes up, the standard error of the sample means goes down.

C.77 The differences in sample proportions will have a mean of $p_A - p_D = 0.170 - 0.159 = 0.011$, and a standard deviation equal to the standard error SE. We have

$$SE = \sqrt{\frac{p_A(1 - p_A)}{n_A} + \frac{p_D(1 - p_D)}{n_D}} = \sqrt{\frac{0.17(0.83)}{200} + \frac{0.159(0.841)}{200}} = 0.037$$

C.79 The sample size is large enough to use the normal approximation. The mean is

$$p_a - p_{ri} = 0.520 - 0.483 = 0.037$$

the standard error is

$$\sqrt{\frac{0.52(1-0.52)}{300} + \frac{0.483(1-0.483)}{300}} = 0.041$$

The distribution of $\hat{p}_a - \hat{p}_{ri}$ is approximately $N(0.037, 0.041)$.

C.81 If we let μ_E represent the average time spent in freeze mode for rats who have been previously shocked (the experimental condition) and μ_C represent the same thing for rats who have not been previously shocked (the control condition), then the hypotheses are

$$H_0: \quad \mu_E = \mu_C$$
$$H_a: \quad \mu_E > \mu_C$$

The sample sizes are small, but the data has no large outliers so we use the t-distribution. The test statistic is

$$t = \frac{\overline{x}_E - \overline{x}_C}{\sqrt{\frac{s_E^2}{n_E} + \frac{s_C^2}{n_C}}} = \frac{36.6 - 1.2}{\sqrt{\frac{21.3^2}{15} + \frac{2.3^2}{11}}} = 6.39$$

This is an upper-tail test, so the p-value is the area above 6.39 in a t-distribution with $df = 10$. We see that this is approximately zero. There is strong evidence that rats with previous experience getting shocked react more strongly to seeing other rats get shocked.

C.83   (a) If p is the proportion of overtime games that are won by the coin flip winner, we test $H_0 : p = 0.5$ vs $H_a : p > 0.5$ to see if there is an advantage. The proportion in the sample is $\hat{p} = 240/428 = 0.561$ and the sample size is large so we compute a standardized z-statistic

$$z = \frac{\hat{p} - p_0}{\sqrt{\frac{p_0(1-p_0)}{n}}} = \frac{0.561 - 0.5}{\sqrt{\frac{0.5(1-0.5)}{428}}} = 2.52$$

Using the upper tail of a standard normal distribution beyond $z = 2.52$, the p-value for this test is 0.006. This is a small p-value, meaning it is quite unlikely to see this many wins by the coin flip winner (if there were no advantage), so we reject the null hypothesis and conclude that there probably is some advantage to winning the coin flip when playing in overtime in the NFL.

(b) To compare the proportions of overtime wins by coin flip winners under the two rules we test $H_0 : p_1 = p_2$ vs $H_a : p_1 \neq p_2$, where p_1 is the proportion under the old rule and p_2 is using the new rule. From the sample data we estimate a proportion of each group and the combined sample.

$$\hat{p}_H = \frac{94}{188} = 0.500 \qquad \hat{p}_A = \frac{146}{240} = 0.608 \qquad \hat{p} = \frac{94 + 146}{188 + 240} = \frac{240}{428} = 0.561$$

The standardized test statistic is

$$z = \frac{0.500 - 0.608}{\sqrt{0.561(1 - 0.561)\left(\frac{1}{188} + \frac{1}{240}\right)}} = \frac{-0.108}{0.04833} = -2.23$$

Since the sample sizes are large we find a p-value using the area in a N(0,1) distribution that lies below $z = -2.23$ and double to account for two tails. This gives p-value $= 2(0.0129) = 0.0258$ which is less than a 5% significance level. This provides evidence that the advantage to the coin flip winner is different under the two sets of rules. However, we should take care to avoid making a cause/effect conclusion about this relationship. These data were not from an experiment (can you imagine randomly assigning a rule to each overtime game?). It is quite possible that some other aspect of the game might have changed between the these two eras that is responsible for the change in proportions for coin flip winners.

C.85 There are 48 people who are lying and the lie detector accurately detected the lies for 31 of them, so we have $\hat{p} = 31/48 = 0.646$. For a 90% confidence interval, we use $z^* = 1.645$ so the confidence interval is

$$\hat{p} \quad \pm \quad z^* \cdot \sqrt{\frac{\hat{p}(1 - \hat{p})}{n}}$$

$$0.646 \quad \pm \quad 1.645 \cdot \sqrt{\frac{0.646(1 - 0.646)}{48}}$$

$$0.646 \quad \pm \quad 0.114$$

$$0.532 \quad \text{to} \quad 0.760.$$

We are 90% confident that a lie detector will accurately detect a lying person under these circumstances between 53.2% and 76.0% of the time.

C.87 This is a hypothesis test for a difference in proportions. Using p_L to represent the proportion found lying when they are lying and p_T to represent the proportion found lying when they are telling the truth, the hypotheses are

$$H_0: \quad p_L = p_T$$
$$H_a: \quad p_L \neq p_T$$

We need three different sample proportions. We have

$$\hat{p}_L \quad = \quad 31/48 = 0.646$$
$$\hat{p}_T \quad = \quad 27/48 = 0.563$$
$$\hat{p} \quad = \quad 58/96 = 0.604$$

The test statistic is

$$z = \frac{\hat{p}_L - \hat{p}_T}{\sqrt{\hat{p}(1 - \hat{p})\left(\frac{1}{n_L} + \frac{1}{n_T}\right)}} = \frac{0.646 - 0.563}{\sqrt{0.604(0.396)\left(\frac{1}{48} + \frac{1}{48}\right)}} = 0.83.$$

This is a two-tail test, so the p-value is twice the area above 0.83 in a normal distribution. We see that the p-value is $2(0.203) = 0.406$. For any reasonable significance level, this is not significant. We do not reject H_0 and do not find evidence that there is any difference in the proportion the machine says are lying depending on whether the person is actually lying or telling the truth.

C.89 We are estimating p, the proportion of US adults who believe the government does not provide enough support for soldiers returning from Iraq or Afghanistan. We have $\hat{p} = 931/1502 = 0.620$ with $n = 1502$. For a 99% confidence interval, we have $z^* = 2.576$. The 99% confidence interval for a proportion is

$$\hat{p} \quad \pm \quad z^* \cdot \sqrt{\frac{\hat{p}(1 - \hat{p})}{n}}$$

$$0.620 \quad \pm \quad 2.576 \cdot \sqrt{\frac{0.620(1 - 0.620)}{1502}}$$

$$0.620 \quad \pm \quad 0.032$$

$$0.588 \quad \text{to} \quad 0.652$$

We are 99% confident that the proportion of US adults who believe the government does not provide enough support for returning soldiers is between 0.588 and 0.652.

C.91 (a) We are estimating $p_s - p_{ns}$ where p_s is the proportion of smokers who get pregnant in the first cycle of trying and p_{ns} is the proportion of non-smokers who get pregnant in the first cycle of trying. For smokers we have $\hat{p}_s = 38/135 = 0.28$. For non-smokers we have $\hat{p}_{ns} = 206/543 = 0.38$. For a 95% confidence interval, we have $z^* = 1.96$. The confidence interval is

$$(\hat{p}_s - \hat{p}_{ns}) \quad \pm \quad z^* \cdot \sqrt{\frac{\hat{p}_s(1 - \hat{p}_s)}{n_s} + \frac{\hat{p}_{ns}(1 - \hat{p}_{ns})}{n_{ns}}}$$

$$(0.28 - 0.38) \quad \pm \quad 1.96 \cdot \sqrt{\frac{0.28(0.72)}{135} + \frac{0.38(0.62)}{543}}$$

$$-0.10 \quad \pm \quad 0.086$$

$$-0.186 \quad \text{to} \quad -0.014$$

We are 95% confident that the difference in pregnancy rates between smoking and non-smoking women during the first cycle of trying to get pregnant is between -0.186 and -0.014. Since 0 (representing no difference) is not in this interval, there appears to be a significant difference between smokers and non-smokers in pregnancy success rates.

(b) The hypotheses are

$$H_0: \quad p_s = p_{ns}$$
$$H_a: \quad p_s \neq p_{ns}$$

In addition to the sample proportions computed in part (a), we need the pooled proportion. We see that out of the 678 women attempting to become pregnant 244 succeeded in their first cycle, so $\hat{p} = 244/678 = 0.36$. The test statistic is

$$z = \frac{\hat{p}_s - \hat{p}_{ns}}{\sqrt{\hat{p}(1 - \hat{p}) \left(\frac{1}{n_s} + \frac{1}{n_{ns}} \right)}} = \frac{0.28 - 0.38}{\sqrt{0.36(0.64) \left(\frac{1}{135} + \frac{1}{543} \right)}} = -2.17$$

This is a two-tail test, so the p-value is twice the area below -2.17 in a standard normal distribution. We see that the p-value is $2(0.015) = 0.030$. At a 5% level, we reject H_0. There is evidence that smokers have less success getting pregnant than non-smokers.

(c) Although the results are significant, the data come from an observational study rather than an experiment. There may be confounding variables and we cannot conclude that there is a causal relationship.

C.93 (a) The sample is the 2,006 randomly selected US adults. The intended population is all US adults.

(b) This is an observational study and we can not make causal conclusions from this study.

(c) This is a hypothesis test for a difference in means. Using μ_S for the mean number of close confidants for those using a social networking site and μ_N for the average number for those not using a social networking site, the hypotheses for the test are:

$$H_0: \quad \mu_S = \mu_N$$
$$H_a: \quad \mu_S > \mu_N$$

The relevant statistic for this test is $\bar{x}_S - \bar{x}_N$, the difference in means for the samples. The relevant null parameter is zero, since from the null hypothesis we have $\mu_S - \mu_N = 0$. The t-test statistic is:

$$t = \frac{\text{Sample statistic} - \text{Null parameter}}{SE} = \frac{(\bar{x}_S - \bar{x}_N) - 0}{\sqrt{\frac{s_S^2}{n_S} + \frac{s_N^2}{n_N}}} = \frac{2.5 - 1.9}{\sqrt{\frac{1.4^2}{947} + \frac{1.3^2}{1059}}} = 9.91.$$

This is a right tail test, so using the t-distribution with df = 947 we see that the p-value is essentially zero. (Indeed, the test statistic is almost 10 standard deviations above the mean!) This is an extremely small p-value, so we reject H_0, and conclude that there is very strong evidence that those with a profile on a social networking site tend, on average, to have more close confidants.

(d) There are many possible confounding variables. Remember that a confounding variable is one with a likely association with both variables of interest (in this case, number of close confidants and whether or not the person is on a social networking site.) One possible confounding variable is age, while another is gender, and still another is how socially active and extroverted a person is. Other answers are possible.

C.95 We estimate $\mu_f - \mu_m$ where μ_f represents the mean time spent exercising for females and μ_m represents the mean time spent exercising for males. For females, we have $\bar{x}_f = 6.40$ with $s_f = 4.60$ and $n_f = 10$. For males, we have $\bar{x}_m = 6.81$ with $s_m = 3.83$ and $n_m = 26$. The data are relatively symmetric with no extreme outliers so we use a t-distribution. For a 95% confidence interval with $df = 9$, we have $t^* = 2.26$. The 95% confidence interval is

$$
\begin{aligned}
(\bar{x}_f - \bar{x}_m) \quad &\pm \quad t^* \cdot \sqrt{\frac{s_f^2}{n_f} + \frac{s_m^2}{n_m}} \\
(6.40 - 6.81) \quad &\pm \quad 2.26 \cdot \sqrt{\frac{4.60^2}{10} + \frac{3.83^2}{26}} \\
-0.41 \quad &\pm \quad 3.70 \\
-4.11 \quad &\text{to} \quad 3.29
\end{aligned}
$$

We are 95% confident that senior females average between 4.11 hours less to 3.29 hours more of exercise per week than senior males.

C.97 Letting μ represent the mean pulse rate of ICU patients, we are testing the hypotheses

$$
\begin{aligned}
H_0 : \quad &\mu = 80 \\
H_a : \quad &\mu > 80
\end{aligned}
$$

The test statistic is

$$\frac{\bar{x} - \mu_0}{\frac{s}{\sqrt{n}}} = \frac{98.9 - 80}{26.8/\sqrt{200}} = 9.97$$

Using a t-distribution with 199 df, we see that the p-value is essentially zero. There is strong evidence that ICU patients have a mean pulse rate higher than 80 beats per minute.

C.99 Letting p_F represent the proportion of men getting prostate cancer while taking finasteride and p_C represent the proportion of men getting prostate cancer in the control group taking the placebo, the hypotheses are

$$
\begin{aligned}
H_0 : \quad &p_F = p_C \\
H_a : \quad &p_F < p_C
\end{aligned}
$$

To compute the test statistic, we need three sample proportions. We have $\hat{p}_F = 804/4368 = 0.184$ and $\hat{p}_C = 1145/4692 = 0.244$. To compute the pooled proportion, we see that a total of $804 + 1145 = 1949$ men got prostate cancer out of a total of $4368 + 4692 = 9060$ men in the study. The pooled proportion is

$$\hat{p} = \frac{804 + 1145}{4368 + 4692} = \frac{1949}{9060} = 0.215$$

The test statistic is

$$z = \frac{\hat{p}_F - \hat{p}_C}{\sqrt{\hat{p}(1 - \hat{p})\left(\frac{1}{n_F} + \frac{1}{n_C}\right)}} = \frac{0.184 - 0.244}{\sqrt{0.215(0.785)\left(\frac{1}{4368} + \frac{1}{4692}\right)}} = -6.95$$

This is a lower-tail test, so the p-value is the area below -6.95 in a normal distribution. This is almost seven standard deviations below the mean, so we see that the p-value is essentially zero. There is very strong evidence that men who take finasteride are less likely to develop prostate cancer.

C.101 This is a test for a mean. If we let μ be the average age of honeybee scouts, we are testing $H_0 : \mu = 12$ vs $H_a : \mu > 12$. The standardized test statistic is

$$t = \frac{\bar{x} - \mu_0}{s/\sqrt{n}} = \frac{29.1 - 12}{5.6/\sqrt{50}} = 21.6$$

This t-statistic is very large, so we know that the p-value is essentially zero. There is very strong evidence that scout bees are older, on average, than the general population of all honeybees.

C.103   (a) The sample proportion is $\hat{p} = 16/90 = 0.178$. The standard error is calculated as:

$$SE = \sqrt{\frac{\hat{p}(1 - \hat{p})}{n}} = \sqrt{\frac{0.178(1 - 0.178)}{90}} = 0.04.$$

   (b) Since we have more than 10 cases (16 is the smallest) in each group the Central Limit Theorem applies.

   (c) Using our answers to parts (a) and (b), and the fact that for a 95% confidence interval, we have $z^* = 1.96$, we calculate

$$\hat{p} \pm z^* \cdot SE = 0.178 \pm 1.96(0.04) = 0.178 \pm 0.078 = (0.100, 0.256)$$

We are 95% sure that between 10.0% and 25.6% of houses for sale in these three Mid-Atlantic states are larger than the national mean.

C.105   (a) The sample difference is our best point estimate

$$\hat{p}_M - \hat{p}_C = \frac{16}{90} - \frac{7}{30} = 0.178 - 0.233 = -0.055$$

   (b) The sample sizes are too small for the Central Limit Theorem to apply ($7 < 10$), so we return to the methods of Chapter 3 and use **StatKey** or other technology to create a bootstrap distribution by taking samples of size 90 and 30 with replacement from the original samples of Mid-Atlantic homes for sale and California homes for sale.

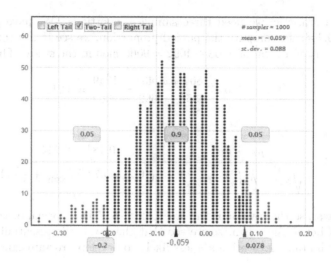

The bootstrap distribution is approximately symmetric. We use the percentile method, keeping the middle 90%, and get an 90% confidence interval of $(-0.200, 0.078)$. We are 90% sure that the proportion of big homes in Mid-Atlantic states is between 0.200 lower and 0.078 higher than the proportion of big homes in California.

(c) Since the 90% confidence interval for the difference in proportions $(-0.200, 0.078)$ contains zero, we do not have clear evidence that the proportion of big homes is different between the two locations, even at a 10% significance level.

C.107 While there are some possible outliers in the 80's age range, they are not extreme and otherwise the data looks reasonable, so we proceed using the t-distribution.

(a) For the 13 teenage patients, we compute $\overline{x}_T = 126.15$ with $s_T = 19.57$. With $df = 12$, we have $t^* = 2.18$. The 95% confidence interval is

$$
\begin{aligned}
\overline{x}_T \;\; &\pm \;\; t^* \cdot \frac{s_T}{\sqrt{n_T}} \\
126.15 \;\; &\pm \;\; 2.18 \cdot \frac{19.57}{\sqrt{13}} \\
126.15 \;\; &\pm \;\; 11.83 \\
114.32 \;\; &\text{to} \;\; 137.98
\end{aligned}
$$

We are 95% confident that the mean systolic blood pressure reading for Intensive Care Unit teenage patients is between 114.32 and 137.98.

For the 15 patients in their eighties, we compute $\overline{x}_E = 132.27$ with $s_E = 31.23$. With $df = 14$, we have $t^* = 2.14$. The 95% confidence interval is

$$
\begin{aligned}
\overline{x}_E \;\; &\pm \;\; t^* \cdot \frac{s_E}{\sqrt{n_E}} \\
132.27 \;\; &\pm \;\; 2.14 \cdot \frac{31.23}{\sqrt{15}} \\
132.27 \;\; &\pm \;\; 17.26 \\
115.01 \;\; &\text{to} \;\; 149.53
\end{aligned}
$$

We are 95% confident that the mean systolic blood pressure reading for Intensive Care Unit patients in their 80s is between 115.01 and 149.53.

The margin of error is larger for the patients in their 80's. Although the sample size is slightly larger for this group, the variability is quite a bit larger, causing the margin of error to be larger.

(b) If μ_T represents the mean systolic blood pressure reading of ICU teenage patients and μ_E represents the mean systolic blood pressure reading for patients in their 80s, the hypotheses are

$$H_0 : \quad \mu_T = \mu_E$$
$$H_a : \quad \mu_T \neq \mu_E$$

Using the summary statistics given in part (a), we compute the test statistic:

$$t = \frac{\bar{x}_T - \bar{x}_E}{\sqrt{\frac{s_T^2}{n_T} + \frac{s_E^2}{n_E}}} = \frac{126.15 - 132.27}{\sqrt{\frac{19.57^2}{13} + \frac{31.23^2}{15}}} = -0.63$$

This is a two-tail test, so the p-value is two times the area below -0.63 in a t-distribution with $df = 12$. We see that the p-value is $2(0.270) = 0.540$. This is a very large p-value so there is no clear evidence of a difference in blood pressure between the two age groups.

C.109 These are paired data so we need to compute the differences, $d = Postest - Pretest$, given below.

$$17.5, \quad 5, \quad 7.5, \quad 15, \quad 5, \quad 25, \quad 15, \quad 17.5, \quad 15, \quad 17.5$$

The mean of these improvement differences is $\bar{x}_d = 14.0$ with a standard deviation of $s_d = 6.37$. The distribution of the differences is relatively symmetric with no big outliers so we use the t-distribution to find the confidence interval for the mean difference, with $t^* = 2.262$ for 95% confidence with $df = 9$.

$$\bar{x}_d \quad \pm \quad t^* \cdot \frac{s_d}{\sqrt{n}}$$
$$14.0 \quad \pm \quad 2.262 \cdot \frac{6.37}{\sqrt{10}}$$
$$14.0 \quad \pm \quad 4.56$$
$$9.44 \quad \text{to} \quad 18.56$$

Based on these data we are 95% sure that the mean improvement between the CAOS posttest and pretest scores for this instructor's students is between 9.44 and 18.56 points.

C.111 The sample includes 240 females and 260 males, both large enough that we needn't be concerned about problems due to lack of normality. The estimated mean commuting times are quite close, $\bar{x}_f = 21.6$ minutes for women and $\bar{x}_m = 22.3$ minutes for men, giving a difference of just 0.7 minutes longer (on average) for men's commutes. If we test for a difference in mean commute time with $H_0 : \mu_f = \mu_m$ vs $H_a : \mu_f \neq \mu_m$ we find a lack of evidence for much difference in the means (p-value=0.585). Based on the confidence interval, we are 95% sure that the mean female commuters in St. Louis average between 3.2 minutes less and 1.8 minute more in commute time compared to males.

C.113 This question suggest a hypothesis test to compare the means commute times between trips made with the carbon and steel bikes. The relevant hypotheses are $H_0 : \mu_c = \mu_s$ vs $H_a : \mu_c \neq \mu_s$, where μ_c and μ_s are the mean commute times (in minutes) for the carbon and steel bikes.

Using technology and the data in the *Minutes* variable of **BikeCommute** we find the summary statistics below for the samples using each type of bike along with boxplots to compare the distributions.

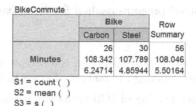

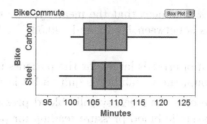

The distributions of commute times in both samples are relatively symmetric and show no outliers, so we use a two-sample t-test with $26 - 1 = 25$ degrees of freedom to compare the means. The t-test statistic is

$$t = \frac{\overline{x}_c - \overline{x}_s}{\sqrt{\frac{s_c^2}{n_c} + \frac{s_s^2}{n_s}}} = \frac{108.34 - 107.79}{\sqrt{\frac{6.25^2}{26} + \frac{4.86^2}{30}}} = 0.36$$

For the two-tailed alternative, we double the area beyond 0.36 for a t-distribution with 25 degrees of freedom to get a p-value= $2(0.3609) = 0.7218$. This is not at all a small p-value, so we fail to find evidence that the mean commute time differs between the two types of bikes. The difference in mean times for these samples was about 0.55 minutes (or about 33 seconds) – and the older, cheaper, steel bike had the smaller (sample) mean time.

C.115   (a) Using the data in **MarriageAges** we find that the mean age at marriage for the 105 wives is $\overline{x}_w = 31.8$ years old with an interval that says we are 95% sure the mean age for wives in the population is between 29.8 and 33.9 years old.

  (b) For the husbands the mean age on marriage licenses in the sample is $\overline{x}_h = 34.7$ years old with a 95% confidence interval for the mean age of husbands in the population going from 32.3 to 37.0 years old.

  (c) You might be tempted to use the fact that the 95% confidence intervals for wives (29.8,33.9) and husbands (32.3,37.0) overlap to conclude that the difference between the mean ages is not significant. This would be wrong on two counts. First, the separate intervals do not take into account the pairing of husbands and wives in the data. Second, just because there is some overlap at the ends of the two intervals, we cannot conclude a common mean is simultaneously plausible for both groups. In fact, a paired data t-test for these data give strong evidence that the mean age is higher for husbands.

C.117 The relevant hypotheses are $H_0 : \rho = 0$ vs $H_a : \rho > 0$, where ρ is the correlation between uniform malevolence and penalty minutes. We can construct randomization samples under this null hypothesis by randomly assigning the malevolence ratings to the standardized penalty minute values. Finding the correlations for 1000 of these randomization samples produces a dotplot such as the one shown below.

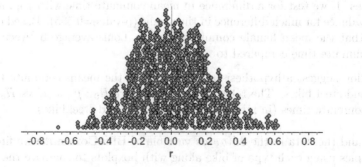

The standard deviation of the correlations in this randomization distribution is $SE = 0.222$. Using the correlation from the original sample of $r = 0.521$ we obtain a standardized test statistic

$$z = \frac{0.521 - 0}{0.222} = 2.35$$

The area above 2.35 in a $N(0, 1)$ distribution, the upper tail p-value, is 0.0094. Since this value is quite small (less than 5%) we have strong evidence that there is a positive correlation between perceived uniform malevolence and penalty minutes for NHL teams.

C.119 (a) The standard deviation of the 5000 *Phat* values in **RandomP50N200** is 0.0354.

(b) The sample proportion is $\hat{p} = 84/200 = 0.42$. The standardized test statistic is

$$z = \frac{0.42 - 0.5}{0.0354} = -2.26$$

(c) If $H_0 : p = 0.5$ is true, we expect (on average) half of the 200 spins, or a count of 100 to be heads.

(d) The standard deviation of the 5000 *Count* values in **RandomP50N200** is 7.081.

(e) We see that

$$z = \frac{84 - 100}{7.081} = -2.26$$

. It is the same as that in (b).

(f) Using technology, the area below -2.26 in a standard normal distribution is 0.012. Since this is a two-tailed test, we double that area to find p-value=$2 \cdot 0.012 = 0.024$. This is a small value, so we have fairly strong evidence that the proportion of heads when spinning a penny differs from 0.50. Therefore penny spinning is probably not a fair process.

Section 7.1 Solutions

7.1 The expected count in each category is $n \cdot p_i = 500(0.25) = 125$. See the table.

Category	1	2	3	4
Expected count	125	125	125	125

7.3 The expected count in category A is $n \cdot p_A = 200(0.50) = 100$. The expected count in category B is $n \cdot p_B = 200(0.25) = 50$. The expected count in category C is $n \cdot p_C = 200(0.25) = 50$. See the table.

Category	A	B	C
Expected count	100	50	50

7.5 We calculate the chi-square statistic using the observed and expected counts.

$$
\begin{aligned}
\chi^2 &= \sum \frac{(observed - expected)^2}{expected} \\
&= \frac{(35-40)^2}{40} + \frac{(32-40)^2}{40} + \frac{(53-40)^2}{40} \\
&= 0.625 + 1.6 + 4.225 \\
&= 6.45
\end{aligned}
$$

There are three categories (A, B, and C) for this categorical variable, so we use a chi-square distribution with degrees of freedom equal to 2. The p-value is the area in the upper tail, which we see is 0.0398.

7.7 We calculate the chi-square statistic using the observed and expected counts.

$$
\begin{aligned}
\chi^2 &= \sum \frac{(observed - expected)^2}{expected} \\
&= \frac{(132-160)^2}{160} + \frac{(181-160)^2}{160} + \frac{(45-40)^2}{40} + \frac{(42-40)^2}{40} \\
&= 4.90 + 2.76 + 0.63 + 0.10 \\
&= 8.38
\end{aligned}
$$

There are four categories (A, B, C, and D) for this categorical variable, so we use a chi-square distribution with degrees of freedom equal to 3. The p-value is the area in the upper tail, which we see is 0.039.

7.9 (a) The sample size is $n = 160$ and the hypothesized proportion is $p_b = 0.25$, so the expected count is $n \cdot p_b = 160(0.25) = 40$.

(b) For the "B" cell we have

$$
\frac{(observed - expected)^2}{expected} = \frac{(36-40)^2}{40} = 0.4
$$

(c) The table has $k = 4$ cells, so the chi-square distribution has $4 - 1 = 3$ degrees of freedom.

7.11 (a) Add the counts in the table to find the sample size is $n = 210 + 732 + 396 + 125 + 213 + 324 = 2000$. The hypothesized proportion is $p_b = 0.35$, so the expected count is $n \cdot p_b = 2000(0.35) = 700$.

(b) For the "B" cell we have

$$
\frac{(observed - expected)^2}{expected} = \frac{(732-700)^2}{700} = 1.46
$$

(c) The table has $k = 6$ cells, so the chi-square distribution has $6 - 1 = 5$ degrees of freedom.

7.13 (a) Let p_g, p_o, p_p, p_r, and p_y be the proportion of people who choose each of the respective flavors. If all flavors are equally popular (1/5 each) the hypotheses are

$$H_0: \quad p_g = p_o = p_p = p_r = p_y = 0.2$$
$$H_a: \quad \text{Some } p_i \neq 0.2$$

(b) If they were equally popular we would have $66(1/5) = 13.2$ people in each category.

(c) Since we have 5 categories we have 4 degrees of freedom.

(d) We calculate the test statistic

$$\chi^2 = \frac{(18 - 13.2)^2}{13.2} + \frac{(9 - 13.2)^2}{13.2} + \frac{(15 - 13.2)^2}{13.2} + \frac{(13 - 13.2)^2}{13.2} + \frac{(11 - 13.2)^2}{13.2}$$
$$= 1.75 + 1.34 + 0.25 + 0.00 + 0.37$$
$$= 3.70$$

(e) The test statistic 3.70 compared to a chi-square distribution with 4 degrees of freedom yields a p-value of 0.449. We fail to reject the null hypothesis, meaning the data doesn't provide significant evidence that some skittle flavors are more popular than others.

7.15 (a) Since there are five groups and we are assuming the groups are equally likely, the proportion in each group should be 1/5 or 0.2. The hypotheses for the test are:

$$H_0: \quad p_a = p_2 = p_3 = p_4 = p_5 = 0.2$$
$$H_a: \quad \text{Some } p_i \text{ is not } 0.2$$

where p_1 represents the proportion of social networking site users in the 18-22 age group, p_2 represents the proportion in the 23-35 age group, and so on. We have $n = 975$ and $p_i = 0.2$ for each group, so the expected count if all age groups are equally likely is $975(0.2) = 195$. The expected counts are all 195, as shown in the table.

Age	18-22	23-35	36-49	50-65	65+
Expected count	195	195	195	195	195

We calculate the chi-square statistic using the observed and expected counts.

$$\chi^2 = \sum \frac{(observed - expected)^2}{expected}$$
$$= \frac{(156 - 195)^2}{195} + \frac{(312 - 195)^2}{195} + \frac{(253 - 195)^2}{195} + \frac{(195 - 195)^2}{195} + \frac{(59 - 195)^2}{195}$$
$$= 7.8 + 70.2 + 17.3 + 0 + 94.9$$
$$= 190.2$$

There are five age groups, so we use a chi-square distribution with degrees of freedom equal to 4. The p-value is the area in the upper tail, which we see is essentially zero. There is strong evidence that users of social networking sites are not equally distributed among these age groups.

(b) The largest contributor to the sum for the chi-square test statistic is 94.9 from the 65+ age group, where the observed count is significantly below the expected count. Many fewer senior citizens use a social networking site than we would expect if users were evenly distributed by age.

7.17 This is a chi-square goodness-of-fit test.

(a) We see that the number of boys diagnosed with ADHD is $6880 + 7982 + 9161 + 8945 = 32,968$.

(b) The expected count for January to March is $n \cdot p_i = 32,968(0.244) = 8044.2$. We find the other expected counts similarly, shown in the table below.

Birth Date	Observed	Expected	Contribution to χ^2
Jan-Mar	6880	8044.2	168.5
Apr-Jun	7982	8505.7	32.2
Jul-Sep	9161	8472.8	55.9
Oct-Dec	8945	7945.3	125.8

(c) The contribution to the chi-square statistic for the January to March cell is

$$\frac{(observed - expected)^2}{expected} = \frac{(6880 - 8044.2)^2}{8044.2} = 168.5$$

This number and the contribution for each of the other cells, computed similarly, are shown in the table above, and the chi-square statistic is the sum: $\chi^2 = 168.5 + 32.2 + 55.9 + 125.8 = 382.4$.

(d) Since there are four categories, one for each quarter of the year, the degrees of freedom is $4 - 1 = 3$. The chi-square test statistic is very large (way out in the far reaches of the tail of the χ^2-distribution) so the p-value is essentially zero.

(e) There is *very* strong evidence that the distribution of ADHD diagnoses for boys differs from the proportions of births in each quarter. By comparing the observed and expected counts we see that younger children in a classroom (Oct-Dec) are diagnosed more frequently than we expect, while older children in a class (Jan-Mar) are diagnosed less frequently.

7.19 Let p_1, p_2, p_3, and p_4 be the proportion of hockey players born in the 1st, 2nd, 3rd, and 4th quarter of the year, respectively. We are testing

$H_0 : p_1 = 0.237, \ p_2 = 0.259, \ p_3 = 0.259$ and $p_4 = 0.245$
H_a: Some p_i is not specified as in H_0

The total sample size is $n = 147 + 110 + 52 + 50 = 359$. The expected counts are $359(0.237) = 85$ for Qtr 1, $359(0.259) = 93$ for Qtr 2, $359(0.259) = 93$ for Qtr 3, and $359(0.245) = 88$ for Qtr 4. The chi-square statistic is

$$\chi^2 = \frac{(147 - 85)^2}{85} + \frac{(110 - 93)^2}{93} + \frac{(52 - 93)^2}{93} + \frac{(50 - 88)^2}{88} = 82.6.$$

We use the chi-square distribution with $4 - 1 = 3$ degrees of freedom, which gives a very small p-value that is essentially zero. This is strong evidence that the distribution of birthdates for OHL hockey players differs significantly from the national proportions.

7.21 (a) A χ^2 goodness-of-fit test was most likely done.

(b) Since the results are given as statistically significant, the χ^2-statistic is likely to be large.

(c) Since the results are given as statistically significant, the p-value is likely to be small.

(d) The categorical variable appears to record the number of deaths due to medication errors in different months at hospitals.

(e) The cell giving the number of deaths in July appears to contribute the most to the χ^2- statistic.

(f) In July, the observed count is probably much higher than the expected count.

7.23   (a) There are 6 actors and we are testing for a difference in popularity. The null hypothesis is that each of the proportions is 1/6 while the alternative hypothesis is that at least one of the proportions is not 1/6. The sample size is $98 + 5 + 23 + 9 + 25 + 51 = 211$, so the expected count for each actor is

$$\text{Expected count for each actor } = n \cdot p_i = 211(1/6) = 35.2$$

The chi-square test statistic calculated using the observed data and these expected counts

$$
\begin{aligned}
\chi^2 &= \sum \frac{(observed - expected)^2}{expected} \\
&= \frac{(98 - 35.2)^2}{35.2} + \frac{(5 - 35.2)^2}{35.2} + \frac{(23 - 35.2)^2}{35.2} + \frac{(9 - 35.2)^2}{35.2} + \frac{(25 - 35.2)^2}{35.2} + \frac{(51 - 35.2)^2}{35.2} \\
&= 112.3 + 25.9 + 4.2 + 19.5 + 3.0 + 7.1 \\
&= 172.0
\end{aligned}
$$

This chi-square statistic gives a very small p-value of essentially zero when compared to a chi-square distribution with 5 degrees of freedom. There is strong evidence of a difference in the popularity of the James Bond actors.

(b) If we eliminate one actor, the null hypothesis is that each of the proportions is 1/5 while the alternative hypothesis is that at least one of the proportions is not 1/5. The sample size without the 5 people who selected George Lazenby is $98 + 23 + 9 + 25 + 51 = 206$, so the expected count for each actor is

$$\text{Expected count for each actor } = n \cdot p_i = 206(1/5) = 41.2$$

The chi-square statistic calculated using the observed data (without Lazenby) and these expected counts is

$$
\begin{aligned}
\chi^2 &= \sum \frac{(observed - expected)^2}{expected} \\
&= \frac{(98 - 41.2)^2}{41.2} + \frac{(23 - 41.2)^2}{41.2} + \frac{(9 - 41.2)^2}{41.2} + \frac{(25 - 41.2)^2}{41.2} + \frac{(51 - 41.2)^2}{41.2} \\
&= 78.3 + 8.0 + 25.2 + 6.4 + 2.3 \\
&= 120.2
\end{aligned}
$$

This is still a very large χ^2-statistic and we again have a p-value of essentially zero when it is compared to a chi-square with 4 degrees of freedom. Even with the Lazenby data omitted, we sill find substantial differences in he proportions of fans who choose the different James Bond actors.

(c) No, we should not generalize the results from this online survey to a population of all movie watchers. This poll was a volunteer poll completed by people visiting a James Bond fan site. This is definitely not a random sample of the movie watching population, and could easily be biased. The best inference we could hope for is to generalize to people who visit a James Bond fan website and who participate in online polls.

7.25   (a) The null hypothesis is $H_0 : p_R = p_X = 0.5$ and the alternative hypothesis is that at least one of the proportions is not 0.5. The expected count in each category is $n \cdot p_i = 436(0.5) = 218$. The chi-square statistic is

$$\chi^2 = \frac{(244 - 218)^2}{218} + \frac{(192 - 218)^2}{218} = 3.10 + 3.10 = 6.20$$

Using a χ^2 distribution with $df = 1$, we obtain a p-value of 0.0128. This gives evidence at a 5% level that these two genetic variations are not equally likely.

(b) The null hypothesis is $H_0 : p = 0.5$ and the alternative hypothesis is $H_a : p \neq 0.5$ where p represents the proportion classified R. (Note that the test would give the same results if we used the proportion classified X.) The sample statistic is $\hat{p} = 244/436 = 0.5596$. The test statistic is

$$z = \frac{\hat{p} - p_0}{\sqrt{\frac{p_0(1-p_0)}{n}}} = \frac{0.5596 - 0.5}{\sqrt{\frac{0.5(0.5)}{436}}} = 2.490$$

Using a standard normal distribution, we see that the area above 2.490 is 0.0064. This is a two-tail test, so the p-value is $2(0.0064) = 0.0128$. This gives evidence at a 5% level that these two genetic variations are not equally likely.

(c) The p-values are equal and the conclusions are identical (and the χ^2-statistic is the square of the z-statistic.)

7.27 According to Benford's Law the hypotheses are

$H_0 :$ $p_1 = 0.301$, $p_2 = 0.176$, $p_3 = 0.125$, $p_4 = 0.097$, $p_5 = 0.079$,
         $p_6 = 0.067$, $p_7 = 0.058$, $p_8 = 0.051$, $p_9 = 0.046$
$H_a :$ At least one of the proportions is different from Benford's law

Here is a table of observed counts for the invoices and expected counts using the Benford proportions and a sample size of 7273.

Digit	1	2	3	4	5	6	7	8	9
Observed	2225	1214	881	639	655	532	433	362	332
Expected	2189.4	1280.7	908.7	704.8	575.9	486.9	421.8	372.0	332.8

The value of the chi-square statistic is

$$\chi^2 = \sum \frac{(observed - expected)^2}{expected} = \frac{(2225 - 2189.4)^2}{2189.4} + \cdots + \frac{(332 - 332.8)^2}{332.8} = 26.66$$

We find the p-value $= 0.0008$ using the upper tail beyond 26.66 of a chi-square distribution with $9 - 1 = 8$ degrees of freedom. This is a very small p-value so we have strong evidence that the first digits of these invoices do not follow Benford's law. The biggest contributions to the chi-square statistic come from an unusually large number of entries starting with "5" and too few with "4". Auditors might want to look more carefully at invoices for amounts beginning with the digit "5".

7.29 (a) To test $H_0 : p_0 = p_1 = p_2 = \cdots = p_9 = 0.10$ vs $H_a :$ Some $p_i \neq 0.10$ the expected count in each cell is $np_i = 150(0.1) = 15$. Here is a table of observed counts for the digits in $RND4$.

```
                      Test              Contribution
Category  Observed  Proportion  Expected   to Chi-Sq
0             12       0.1          15      0.60000
1             14       0.1          15      0.06667
2             16       0.1          15      0.06667
3             13       0.1          15      0.26667
4             22       0.1          15      3.26667
5             10       0.1          15      1.66667
```

6	27	0.1	15	9.60000
7	14	0.1	15	0.06667
8	10	0.1	15	1.66667
9	12	0.1	15	0.60000

N	DF	Chi-Sq	P-Value
150	9	17.8667	0.037

The contribution to to the chi-square statistic for each cell is the value of $\dfrac{(observed - expected)^2}{expected}$. The sum of these values gives the chi-square statistic

$$\chi^2 = 0.6000 + 0.0667 + 0.0667 + \cdots + 0.6000 = 17.87$$

We compare this to the upper tail of a chi-square distribution with $10 - 1 = 9$ degrees of freedom to get a p-value of 0.037. This is a small p-value, providing evidence that the last digits are not randomly distributed.

(b) To test the first digits with $H_0 : p_1 = p_1 = p_2 = \cdots = p_9 = 1/9$ vs H_a : Some $p_i \neq 1/9$, the expected count in each cell is $np_i = 150 \cdot 1/9 = 16.67$. Here is a table of observed counts for the digits in $RND1$.

Category	Observed	Test Proportion	Expected	Contribution to Chi-Sq
1	44	0.111111	16.6667	44.8267
2	23	0.111111	16.6667	2.4067
3	16	0.111111	16.6667	0.0267
4	14	0.111111	16.6667	0.4267
5	13	0.111111	16.6667	0.8067
6	10	0.111111	16.6667	2.6667
7	14	0.111111	16.6667	0.4267
8	11	0.111111	16.6667	1.9267
9	5	0.111111	16.6667	8.1667

N	DF	Chi-Sq	P-Value
150	8	61.68	0.000

The value of the chi-square statistic is $\chi^2 = 44.83 + 2.41 + \cdots + 8.17 = 61.68$ We compare this to the upper tail of a chi-square distribution with $9 - 1 = 8$ degrees of freedom to get a p-value≈ 0. This provides very strong evidence that the first digits of these numbers are not chosen at random. Looking at the observed and expected counts we see that there are far more numbers starting with "1" than we would expect by random chance.

Section 7.2 Solutions

7.31 For the (Group 3, Yes) cell we have

$$\text{Expected count} = \frac{\text{Group 3 row total} \cdot \text{Yes column total}}{n} = \frac{100 \cdot 260}{400} = 65$$

The contribution to the χ^2-statistic from the (Group 3, Yes) cell is $\dfrac{(72-65)^2}{65} = 0.754$.

7.33 We need to find the row total for Control, $40 + 50 + 5 + 15 + 10 = 120$, and the column total for Disagree, $15 + 5 = 20$. Adding the counts in all the cells shows that the overall sample size is $n = 240$. For the (Control, Disagree) cell we have

$$\text{Expected count} = \frac{120 \cdot 20}{240} = 10$$

The contribution to the χ^2-statistic from the (Control, Disagree) cell is $\dfrac{(15-10)^2}{10} = 2.5$.

7.35 This is a 3×2 table so we have $(3-1) \cdot (2-1) = 2$ degrees of freedom. Also, if we eliminate the last row and last column (ignoring the totals) there are 2 cells remaining.

7.37 This is a 2×5 table so we have $(2-1) \cdot (5-1) = 4$ degrees of freedom. Also, if we eliminate the last row and last column there are four cells remaining.

7.39 The hypotheses for testing an association between these two categorical variables are

$H_0 :$ *Award* preference is not related to *Gender*
$H_a :$ *Award* preference is related to *Gender*

The table below shows the observed and expected counts for each cell. For example, the expected count for the (Female, Academy) cell is $31 \cdot 169/362 = 14.5$.

	Academy	Nobel	Olympic	Total
Female	20 (14.5)	76 (69.6)	73 (85.0)	169
Male	11 (16.5)	73 (79.4)	109 (97.0)	193
Total	31	149	182	362

The value of the chi-square statistic is

$$
\begin{aligned}
\chi^2 &= \frac{(20-14.5)^2}{14.5} + \frac{(76-69.6)^2}{69.6} + \frac{(73-85.0)^2}{85.0} \\
&\quad + \frac{(11-16.5)^2}{16.5} + \frac{(73-79.4)^2}{79.4} + \frac{(109-97.0)^2}{97.0} \\
&= 2.08 + 0.59 + 1.69 + 1.83 + 0.52 + 1.48 = 8.20
\end{aligned}
$$

Since this is a 2×3 table we use a chi-square distribution with $(3-1)(2-1) = 2$ degrees of freedom. The area beyond $\chi^2 = 8.20$ is 0.017. This is a fairly small p-value, less than 5%, so we have fairly strong evidence that the award preferences tend to differ between male and female students.

7.41 (a) The two-way table of penguin survival vs type tag is shown:

	Metal	Electronic	Total
Survived	10	18	28
Died	40	32	72
Total	50	50	100

(b) The hypothesis are

H_0 : Type of tag is not related to survival

H_a : Type of tag is related to survival

(c) The table below shows the expected counts, obtained for each cell by multiplying the row total by the column total and dividing by $n = 100$.

	Metal	Electronic
Survived	14	14
Died	36	36

(d) We calculate the chi-square test statistic

$$\chi^2 = \frac{(10-14)^2}{14} + \frac{(18-14)^2}{14} + \frac{(40-36)^2}{36} + \frac{(32-36)^2}{36}$$
$$= 1.143 + 1.143 + 0.444 + 0.444$$
$$= 3.174$$

(e) We compare our test statistic of 3.174 from part (c) to a chi-square with 1 degree of freedom to get a p-value of 0.075. At a 5% level, we do not have enough evidence that the type of tag and survival rate of the penguins are related. (Remember, though, that this does not mean that they are *not* related. A larger sample size might show a relationship.)

7.43 This is a chi-square test for association for a 3×2 table. The relevant hypotheses are

H_0 : Painkiller use is not associates with miscarriages

H_a : Painkiller use is associated with miscarriages

We compute expected counts for all 6 cells. For example, for the (NSAIDs, Miscarriage) cell, we have

$$Expected = \frac{145 \cdot 75}{1009} = 10.8$$

Using a similar process in each cell we find the expected counts shown in the table below.

	Miscarriage	No miscarriage
NSAIDs	10.8	64.2
Acetaminophen	24.7	147.3
No painkiller	109.5	652.5

Notice that all expected counts are above 5, so we are comfortable in using the chi-square distribution. We compute the contribution to the χ^2-statistic, $(observed - expected)^2/expected$, for each cell. The results are shown in the next table.

	Miscarriage	No miscarriage
NSAIDs	4.80	0.81
Acetaminophen	0.02	0.00
No painkiller	0.39	0.06

Adding up all of these contributions, we find $\chi^2 = 6.08$. Using the upper tail of a chi-square distribution with $df = 2$, we see that the p-value is 0.0478. This is significant (just barely) at a 5% level, so we find evidence of an association betweens the use of painkillers and the chance of a miscarriage. Notice that almost all of the contribution to the chi-square statistic comes from the fact that the number of miscarriages after using NSAIDs (aspirin or ibuprofen) is particularly high compared to what is expected if the variables are unrelated. Pregnant women might be wise to avoid these painkillers. However, we cannot assume that NSAIDs are causing the miscarriages (although that might be the case), since these data come from an observational study not an experiment. For example, there could easily be some other condition that women treat with aspirin or ibuprofen that increases the chance of miscarriage.

7.45 (a) This question is best answered with a χ^2 goodness-of-fit test, ignoring gender because gender is not mentioned in the question. The hypotheses are

$$H_0 : p_{grades} = p_{sports} = p_{popular} = 1/3$$
$$H_a : \text{At least one } p_i \neq 1/3$$

If all three answers were equally likely, the expected count for each answer would be $478 \times 1/3 = 159.3$. Therefore, our table of observed (expected) counts is as follows:

Grades	Sports	Popular
247 (159.3)	90 (159.3)	141 (159.3)

We compute the χ^2 statistic as

$$\chi^2 = \sum \frac{(observed - expected)^2}{expected} = \frac{(247 - 159.3)^2}{159.3} + \frac{(90 - 159.3)^2}{159.3} + \frac{(141 - 159.3)^2}{159.3} = 80.53$$

Comparing this to a χ^2 distribution with $3 - 1 = 2$ degrees of freedom yields a p-value of approximately 0. There is extremely strong evidence that grades, sports, and popularity are not equally important among middle school students in these school districts.

(b) This is a χ^2 test for association with a 2×3 table. The hypotheses are
H_0: Gender and what students value are not associated
H_a: Gender and what students value are associated

The expected counts, computed with

$$expected = \frac{row\ total \times column\ total}{overall\ total}$$

are given in the table below.

Expected	Grades	Sports	Popular
Boy	117.3	42.7	67.0
Girl	129.7	47.3	74.0

The χ^2 statistic is computed as follows:

$$\chi^2 = \frac{(117-117.3)^2}{117.3} + \frac{(60-42.7^2)}{42.7} + \frac{(50-67)^2}{67} + \frac{(130-129.7)^2}{129.7} + \frac{(30-47.3)^2}{47.3} + \frac{(91-74)^2}{74}$$
$$= 21.56$$

Because all the expected counts are greater than 5, we can compare 21.56 to a χ^2- distribution with $(3-1) \times (2-1) = 2$ degrees of freedom. The resulting p-value is 0.00002. This provides enough evidence to reject the null hypothesis and conclude that gender is associated with how students answer what is important to them. At least in these school districts in Michigan, middle school boys and girls have different priorities regarding grades, sports, and popularity.

7.47 (a) Architects had the highest proportion of left-handed people $(26/148 = 0.176)$; Orthopedic Surgeon had the highest proportion of right-handed people $(121/132 = 0.917)$.

(b) The null and alternative hypotheses are H_0: Handedness and career are not associated vs. H_a: Handedness and career are associated. The observed and expected counts are shown in the following table.

	Right	Left	Ambidextrous	Total
Psychiatrist	101 (99.6)	10 (12.5)	7 (5.9)	118
Architect	115 (124.9)	26 (15.6)	7 (7.5)	148
Orthopedic Surgeon	121 (111.4)	5 (13.9)	6 (6.7)	132
Lawyer	83 (88.6)	16 (11.1)	6 (5.3)	105
Dentist	116 (111.4)	10 (13.9)	6 (6.7)	132
Total	536	67	32	635

The value of the chi-square statistic is

$$\chi^2 = \frac{(101-99.6)^2}{99.6} + \frac{(10-12.7)^2}{12.7} + \frac{(7-5.9)^2}{5.9} + \frac{(115-124.9)^2}{124.9} + \frac{(26-15.6)^2}{15.6}$$
$$+\frac{(7-7.5)^2}{7.5} + \frac{(121-111.4)^2}{111.4} + \frac{(5-13.9)^2}{13.9} + \frac{(6-6.7)^2}{6.7} + \frac{(83-88.6)^2}{88.6}$$
$$+\frac{(16-11.1)^2}{11.1} + \frac{(6-5.3)^2}{5.3} + \frac{(116-111.4)^2}{111.4} + \frac{(10-13.9)^2}{13.9} + \frac{(6-6.7)^2}{6.7}$$
$$= 19.0$$

The degrees of freedom are $(5-1)(3-1) = 8$, and the area above 19.0 in the chi-square distribution gives a p-value of 0.015.

(c) At the 5% significance level we can conclude that career choice is associated with handedness, while at the 1% level we cannot conclude that there is an association with handedness for these five professions.

7.49 (a) Under a null hypothesis that age and frequency of status updates are unrelated, the expected count for the "Every day" cell under 18-22 years old is

$$\text{Expected count} = \frac{\text{Row Total} \cdot \text{Column Total}}{n} = \frac{136 \cdot 156}{947} = 22.4$$

(b) For $\chi^2 = 210.9$ with $(5-1)(4-1) = 12$ degrees of freedom, we see that the p-value is essentially zero. We have very strong evidence that age of users is related to the frequency of status updates.

7.51 The null hypothesis is that frequency of liking content on Facebook is not different for males and females. The alternative hypothesis is that frequency of liking is related to gender. We compute expected counts for all 10 cells. For example, for the males who like content on Facebook every day, we have

$$Expected = \frac{219 \cdot 386}{877} = 96.4.$$

Computing all the expected counts in the same way, we find the expected counts shown in the following table.

↓Liking/Gender→	Male	Female
Every day	96.4	122.6
3-5 days/week	40.9	52.1
1-2 days/week	57.7	73.3
Every few weeks	37.9	48.1
Less often	153.2	194.8

Notice that all expected counts are over 5 so we can use a chi-square distribution. We compute the contribution to the χ^2-statistic, $(observed - expected)^2/expected$ for each cell and show the results in the next table.

↓Liking/Gender→	Male	Female
Every day	3.90	3.07
3-5 days/week	0.09	0.07
1-2 days/week	0.34	0.26
Every few weeks	0.45	0.36
Less often	1.07	0.84

Adding up all of these contributions, we obtain the χ^2 statistic 10.45. Using a chi-square distribution with $df = (5 - 1) \cdot (2 - 1) = 4$, we see that the p-value is 0.034. At a 5% level, there is evidence of an association between gender and frequency of liking content on Facebook. Females appear more likely to like content at least once a day. The results are significant at the 5% level but not at the 1% level.

7.53 (a) We see in the Total column that 194 endurance athletes were included in the study.

(b) The expected count for sprinters with the R allele is 61.70 and the contribution to the chi-square statistic is 3.792. We find the expected count using

$$\frac{\text{Sprinter row total} \cdot R \text{ column total}}{\text{Sample size}} = \frac{107 \cdot 425}{737} = 61.70$$

The contribution to the chi-square statistic is

$$\frac{(Observed - Expected)^2}{Expected} = \frac{(77 - 61.70)^2}{61.70} = 3.794$$

which is the same (up to round-off) as the computer output.

(c) We see in the bottom row of the computer output that "DF = 2". Since the two-way table has 3 rows and 2 columns, we have $df = (3 - 1) \cdot (2 - 1) = 2$, as given.

(d) We see in the bottom row of the output that the chi-square test statistic is 10.785 and the p-value is 0.005. There is strong evidence that the distribution of alleles for this gene is different between sprinters, endurance athletes, and non-athletes.

(e) The (Sprint, X) cell contributes the most, 5.166, to the χ^2-statistic. The observed count (30) is substantially less than the expected count (45.30). Sprinters are less likely to have this X allele.

(f) The allele R is most over-represented in sprinters (77 observed compared to 61.7 expected). For endurance athletes the most over-represented allele is X (90 observed compared to 82.1 expected).

7.55 (a) We would write "relapse" or "no relapse" on each card. There would be 38 relapse cards and 24 no relapse cards. We would shuffle the cards together and then deal them into three equal piles, signifying the three different groups, desipramine, lithium, and placebo.

(b) Because the p-value is 0.005, about 5 out of 1000 randomization statistics will be greater than or equal to the observed statistic.

7.57 The null hypothesis is that the two variables are related and the alternative hypothesis is that the two variables are not related. The output from one statistics package (Minitab) is given. Output from other packages may look different but will give the same (or similar) chi-square statistic and p-value. We see that the p-value is 0.028 so there is a significant association between these two variables at the 5% level although not at the 1% level. The largest contribution to the chi-square statistic is from the males who do not smoke, and we see that males are less likely to be nonsmokers than expected (which means they are more likely to smoke).

```
Rows: Gender    Columns: PriorSmoke

                  1        2        3

Female          144       93       36
               136.07    99.67    37.27
               0.4626   0.4459   0.0431

Male             13       22        7
                20.93    15.33     5.73
               3.0066   2.8986   0.2798


Cell Contents:        Count
                      Expected count
                      Contribution to Chi-square

Pearson Chi-Square = 7.137, DF = 2, P-Value = 0.028
```

Section 8.1 Solutions

8.1 Both datasets have the same group means of $\overline{x}_1 = 15$ and $\overline{x}_2 = 20$. However, there is so much variability within the groups for Dataset A that we really can't be sure of a difference. In Dataset B, since there is so much less variability between the groups, it seems obvious that the groups come from different populations. (Think of it this way: if the next number we saw was a 20, would we know which group it belongs to in Dataset A? No. Would we know which group it belongs to in Dataset B? Yes!) We have more convincing evidence for a difference between group 1 and group 2 in Dataset B, since the variability within groups is so much less.

8.3 The scales are the same and it appears that the variability is about the same for both datasets. However the means appear to be much farther apart in Dataset A, so we have more convincing evidence for a difference in population means in Dataset A.

8.5 The scales are the same and it appears that the sample means (about 15 and 25) are the same for the two datasets. However, there is much more variability within the groups in Dataset A, so have more convincing evidence for a difference in population means in Dataset B.

8.7 Since there are three groups, the degrees of freedom for the groups is $3 - 1 = 2$. Since the total number of data values is 15, the total degrees of freedom is $15 - 1 = 14$. This leaves 12 degrees of freedom for the Error (or use $n - \#groups = 15 - 3 = 12$). The Mean Squares are found by dividing each SS by its df, so we compute

$$MSG = 120/2 = 60 \quad \text{(for Groups)} \quad \text{and} \quad MSE = 282/12 = 23.5 \quad \text{(for Error)}$$

The F-statistic is the ratio of the mean square values, we have

$$F = \frac{MSG}{MSE} = \frac{60}{23.5} = 2.55$$

The completed table is shown below.

Source	df	SS	MS	F-statistic
Groups	2	120	60	2.55
Error	12	282	23.5	
Total	14	402		

8.9 Since there are three groups, degrees of freedom for the groups is $3 - 1 = 2$. Since the total number of data values is $10 + 8 + 11 = 29$, the total degrees of freedom is $29 - 1 = 28$. This leaves 26 degrees of freedom for the Error (or use $n - \#groups = 29 - 3 = 26$). We also find the missing sum of squares for Error by subtraction, $SSE = SSTotal - SSG = 1380 - 80 = 1300$. The Mean Squares are found by dividing each SS by its df, so we compute

$$MSG = 80/2 = 40 \quad \text{(for Groups)} \quad \text{and} \quad MSE = 1300/26 = 50 \quad \text{(for Error)}$$

The F-statistic is the ratio of the mean square values, we have

$$F = \frac{MSG}{MSE} = \frac{40}{50} = 0.8$$

The completed table is shown below.

Source	df	SS	MS	F-statistic
Groups	2	80	40	0.8
Error	26	1300	50	
Total	28	1380		

8.11 (a) Since the degrees of freedom for the groups is 3, the number of groups is 4.

(b) The hypotheses are

$$H_0 : \quad \mu_1 = \mu_2 = \mu_3 = \mu_4$$
$$H_a : \quad \text{Some } \mu_i \neq \mu_j$$

(c) Using 3 and 16 for the degrees of freedom with the F-distribution, we see the upper-tail area beyond $F=1.60$ gives a p-value of 0.229.

(d) We do not reject H_0. We do not find convincing evidence for any differences between the population means.

8.13 (a) Since the degrees of freedom for the groups is 2, the number of groups is 3.

(b) The hypotheses are

$$H_0 : \quad \mu_1 = \mu_2 = \mu_3$$
$$H_a : \quad \text{Some } \mu_i \neq \mu_j$$

(c) Using 2 and 27 for the degrees of freedom with the F-distribution, we see the upper-tail area beyond $F=8.60$ gives a p-value of 0.0013.

(d) We reject H_0. We find strong evidence for differences among the population means.

8.15 (a) One variable is which group the girl is in, which is categorical. The other variable is the change in cortisol level, which is quantitative.

(b) This is an experiment, since the researchers assigned the girls to the different groups.

(c) The hypotheses are

$$H_0 : \quad \mu_1 = \mu_2 = \mu_3 = \mu_4$$
$$H_a : \quad \text{Some } \mu_i \neq \mu_j$$

where the four means represent the mean cortisol change after a stressful event for girls who talk to their mothers in person, who talk to their mothers on the phone, who text their mothers, and who have no contact with their mothers, respectively.

(d) Since the overall sample size is 68, total degrees of freedom is 67. Since there are four groups, the df for groups is 3. This leaves 64 degrees of freedom for the error (or use $68 - 4 = 64$).

(e) Since they found a significant difference in mean cortisol change between at least two of the groups, the F-statistic must be significant, which means its p-value is less than 0.05.

8.17   (a) Using μ_r, μ_g, and μ_b to represent mean number of anagrams solved by someone with prior exposure to red, green, and black, respectively, the hypotheses are

$$H_0: \qquad \mu_r = \mu_g = \mu_b$$
$$H_a: \qquad \text{At least two of the means are different}$$

(b) We subtract to see that $SSE = SStotal - SSG = 84.7 - 27.7 = 57.0$. Since there are three groups, df for color is $3 - 1 = 2$. Since the total sample size is $19 + 27 + 25 = 71$, total df is $71 - 1 = 70$. Thus, the error df is 68. We divide by the respective degrees of freedom to find MSG and MSE and then divide those to find the F-statistic. The analysis of variance table is shown, and the F-statistic is 16.5.

```
Source    DF    SS     MS      F
Groups    2     27.7   13.85   16.5
Error     68    57.0   0.84
Total     70    84.7
```

(c) We find the area above $F = 16.5$ in an F-distribution with numerator df equal to 2 and denominator df equal to 68, and see that the p-value is essentially zero.

(d) Reject H_0 and conclude that the means are not all the same. The color of prior instructions has an effect on how students perform on this anagram test. By looking at the sample group means, it appears that seeing red prior to the test may hinder students ability to solve the anagrams.

8.19   (a) Yes, the control groups spend less time in darkness than the groups that were stressed. The mean time in darkness is smaller for each of the groups with "HC" than the means for any of the stressed groups with "SD"". However, of the groups that were stressed, the mice that spent time in an enriched environment (EE) appear to spend less time (on average) in darkness than the other two stressed groups.

(b) The null hypothesis is that environment and prior stress do not affect mean amount of time in darkness, while the alternative is that environment and prior stress do affect mean amount of time in darkness. To construct the ANOVA table we compute

$$MSG = \frac{482776}{5} = 96355.2 \qquad MSE = \frac{177835}{42} = 4234.2 \qquad F = \frac{MSG}{MSE} = \frac{96355.2}{4234.2} = 22.76$$

We summarize these calculations in the ANOVA table below.

```
Source  DF     SS      MS       F      P
Light   5      481776  96355.2  22.76  0.000
Error   42     177835  4234.2
Total   47     659611
```

From an F-distribution with 5 and 42 degrees of freedom, we see that the p-value is essentially zero. There is strong evidence that the average amount of time spent in darkness is not the same for all six combinations of environment and stress. We will see in the next section how to tell where the differences lie.

8.21   (a) In each of the three environments, mice in the SD group who were subjected to stress have lower levels than mice in the HC group, which matches what we expect since stress reduces levels of FosB+ cells. In both the HC (no stress group) and the SD (stress) group, the mice in the enriched environment (EE) have the highest levels of FosB+ cells.

(b) The null hypothesis is that environment and prior stress do not affect mean FosB+ levels, while the alternative is that environment and prior stress do affect these mean levels. To construct the ANOVA table we compute

$$MSG = \frac{118286}{5} = 23657.2 \qquad MSE = \frac{75074}{36} = 2085.4 \qquad F = \frac{MSG}{MSE} = \frac{23657.2}{2085.4} = 11.3$$

We summarize these calculations in the ANOVA table below.

```
Source  DF     SS       MS      F       P
Light    5  118286   23657.2   11.3   0.0000
Error   36   75074    2085.4
Total   41  193360
```

From an F-distribution with 5 and 36 degrees of freedom, we see that the p-value is essentially zero. There is strong evidence that mean FosB+ levels are not the same for all six combinations of environment and stress. We will see in the next section how to tell where the differences lie.

8.23 The null hypothesis is that the mean change in pain threshold is the same regardless of pose struck, $H_0 : \mu_1 = \mu_2 = \mu_3$, while the alternative hypothesis is that mean change in pain threshold is different between at least two of the types of poses, $H_a :$ Some $\mu_i \neq \mu_j$. We find the sum of squares using the shortcut formulas at the end of this section. (The difference between $SSG + SSE$ and $SSTotal$ is due to rounding of the standard deviations.)

$$SSG = 30(14.3 - 1.33)^2 + 29(-4.4 - 1.33)^2 + 30(-6.1 - 1.33)^2 = 7,654.9$$
$$SSE = 29(34.8)^2 + 28(31.9)^2 + 29(35.4)^2 = 99,954.9$$
$$SSTotal = 88(35.0)^2 = 107,800$$

We fill in the rest of the ANOVA table by dividing each sum of square by its respective degrees of freedom and finding the ratio of the mean squares to compute the F-statistic.

$$MSG = \frac{7654.9}{2} = 3827.5 \qquad MSE = \frac{99954.9}{86} = 1162.3 \qquad F = \frac{MSG}{MSE} = \frac{3827.5}{1162.3} = 3.29$$

Using an F-distribution with 2 and 86 degrees of freedom, we find the area beyond $F = 3.29$ gives a p-value of 0.042. We reject H_0 and find evidence that the type of pose a person assumes is associated with change in mean pain threshold. The ANOVA table summarizing these calculations is shown below.

```
Source  DF    SS       MS      F      P
Light    2  7654.9   3827.5   3.29   0.042
Error   86  99954.9  1162.3
Total   88  107800
```

8.25 We are testing $H_0 : \mu_D = \mu_L = \mu_W$ where μ represents the mean number of words recalled. Using the summary statistics we compute the sums of squares needed for the ANOVA table:

$$SSG = 16(4.8 - 3.8)^2 + 16(3.4 - 3.8)^2 + 16(3.2 - 3.8)^2 = 24.32$$
$$SSE = 15(1.3)^2 + 15(1.6)^2 + 15(1.2)^2 = 85.35$$
$$SSTotal = 47(1.527)^2 = 109.59$$

We summarize the calculations in the ANOVA table below. Note: Some values may differ slightly due to roundoff and once you have any two of the sum of squares you can easily find the third.

Source	df	SS	MS	F-value	p-value
Group	2	24.32	12.16	6.40	0.0036
Error	45	85.35	1.90		
Total	47	109.67			

(a) Drawing an image of each word produced the highest mean recall (4.8). Just writing the word down produced the lowest mean recall (3.2).

(b) From the ANOVA table, the F-statistic is $12.16/1.90 = 6.40$.

(c) We see that the area in an F distribution in the right tail beyond 6.40, using 2 and 45 degrees of freedom, is 0.0036. This p-value of 0.0036 is also shown in the ANOVA table.

(d) The p-value is less than 0.05, so we reject H_0. We have strong evidence of a difference in mean number of words recalled depending on whether participants draw an image, list attributes, or write the word.

(e) It does make a difference, and people wanting to memorize words should draw an image!

8.27 We are testing $H_0 : \mu_D = \mu_V = \mu_W$ where μ represents the mean number of words recalled. Using the summary statistics we compute the sums of squares needed for the ANOVA table:

$$\begin{aligned} SSG &= 12(4.4 - 3.6)^2 + 12(3.4 - 3.6)^2 + 12(3.0 - 3.6)^2 = 12.48 \\ SSE &= 11(1.2)^2 + 11(1.5)^2 + 11(1.1)^2 = 53.90 \\ SSTotal &= 35(1.377)^2 = 66.36 \end{aligned}$$

We summarize the calculations in the ANOVA table below. Note: Some values may differ slightly due to roundoff and once you have any two of the sum of squares you can easily find the third.

Source	df	SS	MS	F-value	p-value
Group	2	12.48	6.24	3.83	0.032
Error	33	53.90	1.63		
Total	35	66.38			

(a) Drawing an image of each word produced the highest mean recall (4.4). Just writing the word down produced the lowest mean recall (3.0).

(b) From the ANOVA table, the F-statistic is $6.42/1.63 = 3.83$

(c) We see that the area in an F distribution in the right tail beyond 3.83, using 2 and 33 degrees of freedom, is 0.032. This p-value of 0.032 is also shown in the ANOVA table.

(d) The p-value is less than 0.05, so we reject H_0. We have evidence of a difference in mean number of words recalled depending on whether participants draw an image, view an image, or write the word.

(e) It does make a difference, and people wanting to memorize words should draw an image!

8.29 (a) The null hypothesis is that the amount of light at night does not affect how much weight is gained. The alternative hypothesis is that the amount of light at night has some effect on mean weight gain.

(b) We see from the computer output that the F-statistic is 8.38 while the p-value is 0.002. This is a small p-value, so we reject H_0. There is evidence that mean weight gain in mice is influenced by the amount of light at night.

(c) Yes, there is an association between weight gain and light conditions at night. From the means of the groups, it appears that mice with more light at night tend to gain more weight.

(d) Yes, we can conclude that light at night causes weight gain (in mice), since the results are significant and come from a randomized experiment.

8.31 (a) The standard deviations are very different. In particular, the standard deviation for the LL sample ($s_{LL} = 1.31$) is more than double the standard deviation for the LD sample ($s_{LD} = 0.43$). This indicates that an ANOVA test may not be appropriate in this situation.

(b) A p-value of 0.652 from the randomization distribution is not small, so we would not reject a null hypothesis that the means are equal. There is not sufficient evidence to conclude that the mean amount consumed is different depending on the amount of light at night. Mice in different light at night conditions appear to eat the roughly similar amounts on average. Weight gain in mice with light at night is not a result of eating more food.

8.33 (a) For mice in this sample on a standard light/dark cycle, 36.0% of their food is consumed (on average) during the day and the other 64.0% is consumed at night. For mice with even just dim light at night, 55.5% of their food is consumed (on average) during the day and the other 44.5% is consumed at night.

(b) The p-value is 0.000 so there is very strong evidence that light at night influences when food is consumed by mice. Since the result comes from a randomized experiment, and the mice with more light average a higher percentage of food consumed during the day, we can conclude that light at night causes mice to eat a greater percentage of their food during the day. This raises the question: how strong is the association between time of calorie consumption and weight gain? That answer will have to wait until the next chapter.

8.35 (a) The summary statistics are shown below. The controls who had never played football had the largest mean hippocampus volume, and the football players with a history of concussions had the smallest.

```
Group        N   Mean    StDev
Control      25  7602.6  1074.0
FBConcuss    25  5734.6  593.4
FBNoConcuss  25  6459.2  779.7
```

(b) The ANOVA table is shown below. We see that the F-statistic is 31.47 and the p-value is 0.000.

```
Source  DF    SS        MS        F      P
Group   2     44348606  22174303  31.47  0.000
Error   72    50727336  704546
Total   74    95075942
```

(c) This is a very small p-value, so we reject H_0. We have very strong evidence that mean hippocampus size is different depending on one's football playing and concussion experience.

8.37 We start by finding the mean and standard deviations for *Exercise* within each of the award groups and for the whole sample.

	Academy	Nobel	Olympic	Total
Mean	6.81	6.96	11.14	9.05
Std. Dev.	4.09	4.74	5.98	5.74
Sample size	31	148	182	361

The standard deviations are not too different, so the equal variance condition is reasonable. Side-by-side boxplots of exercise between the award groups show some right skewness, but the sample sizes are large so this shouldn't be a problem.

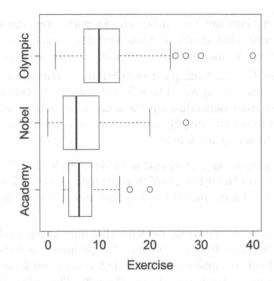

To test $H_0 : \mu_1 = \mu_2 = \mu_3$ vs $H_a :$ Some $\mu_i \neq \mu_j$ we use technology to produce an ANOVA table.

```
            Df   Sum Sq  Mean Sq  F value    Pr(>F)
Award        2   1597.9   798.96   27.861    0.0000
Residuals  358  10266.3    28.68
Total      360  11864.2
```

The p-value from the ANOVA (0.0000) is very small, so we have strong evidence that at least one of the award groups has a mean exercise rate that differs from at least one of the other groups.

8.39 (a) Treatment level appears to be associated with drug resistance, with more aggressive treatments yielding more drug resistance. Treatment level does not appear to be associated with health outcomes.

(b) The p-value from the ANOVA test for ResistanceDensity is ≈ 0, the p-value for DaysInfectious is 0.0002, the p-value for Weight is 0.906, and the p-value for RBC is 0.911. The two response variables measuring drug resistance show a significant difference in means among the dosage groups, while the two response variables measuring health do not show significant differences in means.

(c) Treatment level is significantly associated with drug resistance, and in the sample more aggressive treatments yield *more* drug resistance. This contradicts conventional wisdom that aggressive treatments are better at preventing drug resistance.

(d) With the untreated group excluded, the conditions for ANOVA are most reasonable for the *DaysInfectious* variable. The smallest groups standard deviation (2.48 for the light treatment) is within a factor of two of the largest standard deviation (4.01 for the aggressive treatment). There may be some mild concerns with normality in the boxplot. The other response measuring drug resistance, *ResistanceDensity*, shows a serious concern with the condition on similar variances, since the boxplot shows the variability is much higher in the aggressive treatment group than in the others. For the response variables measuring health, the condition of normality is most obviously violated, because sample sizes are relatively small (only 18 in each group), and there are a couple of extreme outliers in the light and moderate treatment groups for both variables.

Section 8.2 Solutions

8.41 Yes, the p-value (0.002) is very small so we have strong evidence that there is a difference in the population means between at least two of the groups.

8.43 We see in the output that the sample mean for group A is 10.2 with a sample size of 5. For a confidence interval for a mean after an analysis of variance test, we use $\sqrt{MSE}$ for the standard deviation, and we use error df for the degrees of freedom. For a 95% confidence interval with 12 degrees of freedom, we have $t^* = 2.18$. The confidence interval is

$$
\begin{array}{ccc}
\overline{x}_A & \pm & t^* \cdot \dfrac{\sqrt{MSE}}{\sqrt{n_A}} \\[3mm]
10.2 & \pm & 2.18 \cdot \dfrac{\sqrt{6.20}}{\sqrt{5}} \\[3mm]
10.2 & \pm & 2.43 \\[2mm]
7.77 & \text{to} & 12.63
\end{array}
$$

We are 95% confident that the population mean for group A is between 7.77 and 12.63.

8.45 We are testing $H_0 : \mu_A = \mu_C$ vs $H_a : \mu_A \neq \mu_C$. The test statistic is

$$
t = \frac{\overline{x}_A - \overline{x}_C}{\sqrt{MSE\left(\frac{1}{n_A} + \frac{1}{n_C}\right)}} = \frac{10.2 - 10.8}{\sqrt{6.20\left(\frac{1}{5} + \frac{1}{5}\right)}} = -0.38
$$

This is a two-tail test so the p-value is twice the area below -0.38 in a t-distribution with $df = 12$. We see that the p-value is $2(0.355) = 0.71$. This is a very large p-value and we do not find any convincing evidence of a difference in population means between groups A and C.

8.47 The pooled standard deviation is $\sqrt{MSE} = \sqrt{48.3} = 6.95$. We use the error degrees of freedom, which we see in the output is 20.

8.49 We see in the output that the sample mean for group C is 80.0 while the sample mean for group D is 69.33. Both sample sizes are 6. For a confidence interval for a difference in means after an analysis of variance test, we use $\sqrt{MSE}$ for both standard deviations, and we use the error df degrees of freedom. For a 95% confidence interval with 20 degrees of freedom, we have $t^* = 2.09$. The confidence interval is

$$
\begin{array}{ccc}
(\overline{x}_C - \overline{x}_D) & \pm & t^* \cdot \sqrt{MSE\left(\dfrac{1}{n_C} + \dfrac{1}{n_D}\right)} \\[3mm]
(80.00 - 69.33) & \pm & 2.09 \cdot \sqrt{48.3\left(\dfrac{1}{6} + \dfrac{1}{6}\right)} \\[3mm]
10.67 & \pm & 8.39 \\[2mm]
2.28 & \text{to} & 19.06
\end{array}
$$

We are 95% confident that the difference in the population means of group C and group D is between 2.28 and 19.06.

8.51 We are testing $H_0 : \mu_A = \mu_B$ vs $H_a : \mu_A \neq \mu_B$. The test statistic is

$$t = \frac{\overline{x}_A - \overline{x}_B}{\sqrt{MSE\left(\frac{1}{n_A} + \frac{1}{n_B}\right)}} = \frac{86.83 - 76.17}{\sqrt{48.3\left(\frac{1}{6} + \frac{1}{6}\right)}} = 2.66$$

This is a two-tail test so the p-value is twice the area above 2.66 in a t-distribution with $df = 20$. We see that the p-value is $2(0.0075) = 0.015$. With a 5% significance level, we reject H_0 and find evidence of a difference in the population means between groups A and B.

8.53 To compare mean ant counts for peanut butter and ham & pickles the relevant hypotheses are $H_0 : \mu_2 = \mu_3$ vs $H_a : \mu_2 \neq \mu_3$. We compare the sample means, $\overline{x}_2 = 34.0$ and $\overline{x}_3 = 49.25$, and standardize using the SE for a difference in means after ANOVA.

$$t = \frac{\overline{x}_2 - \overline{x}_3}{\sqrt{MSE\left(\frac{1}{n_2} + \frac{1}{n_3}\right)}} = \frac{34.0 - 49.25}{\sqrt{138.7\left(\frac{1}{8} + \frac{1}{8}\right)}} = \frac{-15.25}{5.89} = -2.59$$

We find the p-value using a t-distribution with 21 (error) degrees of freedom, doubling the area below $t = -2.59$ to get p-value $= 2(0.0085) = 0.017$. This is a small p-value so we have evidence of a difference in mean number of ants between peanut butter and ham & pickles sandwiches, with ants seeming to prefer ham & pickles.

8.55 We have three pairs to test. We first test $H_0 : \mu_{DM} = \mu_{LD}$ vs $H_a : \mu_{DM} \neq \mu_{LD}$. The test statistic is

$$t = \frac{\overline{x}_{DM} - \overline{x}_{LD}}{\sqrt{MSE\left(\frac{1}{n_{DM}} + \frac{1}{n_{LD}}\right)}} = \frac{7.859 - 5.987}{\sqrt{6.48\left(\frac{1}{10} + \frac{1}{9}\right)}} = 1.60.$$

This is a two-tail test so the p-value is twice the area above 1.60 in a t-distribution with $df = 25$. We see that the p-value is $2(0.061) = 0.122$. We don't find convincing evidence for a difference in mean weight gain between the dim light condition and the light/dark condition.

We next test $H_0 : \mu_{DM} = \mu_{LL}$ vs $H_a : \mu_{DM} \neq \mu_{LL}$. The test statistic is

$$t = \frac{\overline{x}_{DM} - \overline{x}_{LL}}{\sqrt{MSE\left(\frac{1}{n_{DM}} + \frac{1}{n_{LL}}\right)}} = \frac{7.859 - 11.010}{\sqrt{6.48\left(\frac{1}{10} + \frac{1}{9}\right)}} = -2.69.$$

This is a two-tail test so the p-value is twice the area below -2.69 in a t-distribution with $df = 25$. We see that the p-value is $2(0.0063) = 0.0126$. At a 5% level, we do find a difference in mean weight gain between the dim light condition and the bright light condition, with higher mean weight gain in the bright light condition.

Finally, we test $H_0 : \mu_{LD} = \mu_{LL}$ vs $H_a : \mu_{LD} \neq \mu_{LL}$. The test statistic is

$$t = \frac{\overline{x}_{LD} - \overline{x}_{LL}}{\sqrt{MSE\left(\frac{1}{n_{LD}} + \frac{1}{n_{LL}}\right)}} = \frac{5.987 - 11.010}{\sqrt{6.48\left(\frac{1}{9} + \frac{1}{9}\right)}} = -4.19.$$

This is a two-tail test so the p-value is twice the area below -4.19 in a t-distribution with $df = 25$. We see that the p-value is $2(0.00015) = 0.0003$. There is strong evidence of a difference in mean weight gain between the light/dark condition and the bright light condition, with higher mean weight gain in the bright light condition.

8.57 (a) The p-value for the ANOVA table is essentially zero, so we have strong evidence that there are differences in mean time spent in darkness among the six treatment combinations. Since all the sample sizes are the same, the groups most likely to show a difference in population means are the ones with the sample means farthest apart. These groups are IE:HC (impoverished environment with no added stress, $\overline{x}_1 = 192$) and SE:SD (standard environment with added stress, $\overline{x}_5 = 438$). Likewise, the groups least likely to show a difference in population means are the ones with the sample means closest together. These groups are IE:HC (impoverished environment with no added stress, $\overline{x}_1 = 192$) and SE:HC (standard environment with no added stress, $\overline{x}_2 = 196$).

(b) Given the six means, there appear to be two distinct groups. The four means (all environments with no added stress, along with the enriched environment with added stress) are all somewhat similar, while the other two means (impoverished or standard environment with added stress) appear to be much larger than the other four and similar to each other. Only the enriched environment with its opportunities for exercise appears to have conferred some immunity to the added stress.

(c) Let μ_1 and μ_6 be the means for mice in IE:HC and EE:SD conditions, respectively. We test $H_0 : \mu_1 = \mu_6$ vs $H_a : \mu_1 \neq \mu_6$. The test statistic is

$$t = \frac{\overline{x}_1 - \overline{x}_6}{\sqrt{MSE\left(\frac{1}{n_1} + \frac{1}{n_6}\right)}} = \frac{192 - 231}{\sqrt{2469.9\left(\frac{1}{8} + \frac{1}{8}\right)}} = -1.57.$$

This is a two-tail test so the p-value is twice the area below -1.57 in a t-distribution with $df = 42$. We see that the p-value is $2(0.062) = 0.124$. We do not reject H_0, and do not find convincing evidence of a difference in mean time spent in darkness between mice in the two groups. Prior exercise in an enriched environment may help eliminate the effects of the added stress.

8.59 We test $H_0 : \mu_1 = \mu_4$ vs $H_a : \mu_1 \neq \mu_4$ where μ_1 and μ_4 are the mean change in closeness rating after doing a synchronized, high exertion activity (HS+HE) and a non-synchronized, low exertion activity (LS+LE), respectively. From the table of group means we see $\overline{x}_1 = 0.319$ for a sample of $n_1 = 72$ and $\overline{x}_4 = -0.431$ for a sample of $n_4 = 58$. We also find $MSE = 3.248$ in the ANOVA table, so the t-statistic is

$$t = \frac{\overline{x}_1 - \overline{x}_4}{\sqrt{MSE\left(\frac{1}{n_1} + \frac{1}{n_4}\right)}} = \frac{0.319 - (-0.431)}{\sqrt{3.248\left(\frac{1}{72} + \frac{1}{58}\right)}} = 2.36$$

The p-value is twice the area above 2.36 in a t-distribution with $df = 256$. We see that the p-value is $2(0.0095) = 0.019$. This is a small p-value (less than 5%) so we have convincing evidence that the mean change in closeness rating for a synchronized, high exertion activity is higher than for a non-synchronized, low exertion activity. Note that the HS+HE group had the smallest mean of the three activity groups that were similar and positive, so we can draw a similar conclusion about how the other two groups relate to the LS+LE group.

8.61 We have 42 degrees of freedom for the $MSE = 2469.9$ so the t-value for a 95% confidence interval is $t^* = 2.018$. The sample size in each group is 8, so the value of LSD for comparing any two means is

$$LSD = 2.018\sqrt{2469.9\left(\frac{1}{8} + \frac{1}{8}\right)} = 50.1$$

We write the group means down in increasing order

```
Group: IE:HC   SE:HC   EE:HC   EE:SD   IE:SD   SE:SD
Mean:   192     196     205     231     392     438
```

The first four sample means are all within 50.1 of each other so there are no significant differences in mean time in darkness for any of the non-stressed mice groups or the stressed group that has an enriched environment. The mean time is higher for the other two stressed groups (both means are more than 50.1 seconds above the EE:SD group mean), although those two are not significantly different from each other.

Section 9.1 Solutions

9.1 The estimates for β_0 and β_1 are given in two different places in this output, with the output in the table having more digits of accuracy. We see that the estimate for the intercept β_0 is $b_0 = 29.3$ or 29.266 and the estimate for the slope β_1 is $b_0 = 4.30$ or 4.2969. The least squares line is $\widehat{Y} = 29.3 + 4.30X$.

9.3 The estimates for β_0 and β_1 are given in the "Estimate" column of the computer output. We see that the estimate for the intercept β_0 is $b_0 = 77.44$ and the estimate for the slope β_1 is $b_1 = -15.904$. The least squares line is $\widehat{Y} = 77.44 - 15.904 \cdot Score$.

9.5 The slope is $b_1 = -8.1952$. The null and alternative hypotheses for testing the slope are $H_0 : \beta_1 = 0$ vs $H_a : \beta_1 \neq 0$. We see in the output that the p-value is 0.000, so there is strong evidence that the explanatory variable X is an effective predictor of the response variable Y.

9.7 The slope is $b_1 = -0.3560$. The null and alternative hypotheses for testing the slope are $H_0 : \beta_1 = 0$ vs $H_a : \beta_1 \neq 0$. We see in the output that the p-value is 0.087. At a 5% level, we do not find evidence that the explanatory variable *Dose* is an effective predictor of the response variable for this model.

9.9 A confidence interval for the slope β_1 is given by $b_1 \pm t^* \cdot SE$. For a 95% confidence interval with $df = n - 2 = 22$, we have $t^* = 2.07$. We see from the output that $b_1 = -8.1952$ and the standard error for the slope is $SE = 0.9563$. A 95% confidence interval for the slope is

$$
\begin{array}{ccc}
b_1 & \pm & t^* \cdot SE \\
-8.1952 & \pm & 2.07(0.9563) \\
-8.1952 & \pm & 1.980 \\
-10.1752 & \text{to} & -6.2152
\end{array}
$$

We are 95% confident that the population slope β_1 for this model is between -10.18 and -6.22. We can't give a more informative interpretation without some context for the variables.

9.11 We are testing $H_0 : \rho = 0$ vs $H_a : \rho > 0$. To find the test statistic, we use

$$
t = \frac{r\sqrt{n-2}}{\sqrt{1-r^2}} = \frac{0.35\sqrt{28}}{\sqrt{1-(0.35)^2}} = 1.98
$$

This is a one-tail test, so the p-value is the area above 1.98 in a t-distribution with $df = n - 2 = 28$. We see that the p-value is 0.029. At a 5% level, we have sufficient evidence for a positive linear association between the two variables.

9.13 We are testing $H_0 : \rho = 0$ vs $H_a : \rho \neq 0$. To find the test statistic, we use

$$
t = \frac{r\sqrt{n-2}}{\sqrt{1-r^2}} = \frac{0.28\sqrt{98}}{\sqrt{1-(0.28)^2}} = 2.89
$$

This is a two-tail test, so the p-value is twice the area above 2.89 in a t-distribution with $df = n - 2 = 98$. We see that the p-value is $2(0.0024) = 0.0048$. We find strong evidence of a linear association between the two variables.

9.15　(a) The two variables most strongly positively correlated are *Height* and *Weight*. The correlation is $r = 0.619$ and the p-value is 0.000 to three decimal places. A positive correlation in this context means that taller people tend to weigh more.

(b) The two variables most strongly negatively correlated are GPA and $Weight$. The correlation is $r = -0.217$ and the p-value is 0.000 to three decimal places. A negative correlation in this context means that heavier people tend to have lower grade point averages.

(c) At a 5% significance level, almost all pairs of variables have a significant correlation. The only pair not significantly correlated is hours of $Exercise$ and hours of TV with $r = 0.010$ and p-value=0.852, Based on these data, we find no convincing evidence of a linear association between time spent exercising and time spent watching TV.

9.17 (a) On the scatterplot, we have concerns if there is a curved pattern (there isn't) or variability from the line increasing or decreasing in a consistent way (it isn't) or extreme outliers (there aren't any). We do not have any strong concerns about using these data to fit a linear model.

(b) Using the fitted least squares line in the output, for a student with a verbal SAT score of 650, we have

$$\widehat{GPA} = 2.03 + 0.00189 VerbalSAT = 2.03 + 0.00189(650) = 3.2585$$

The model predicts that a student with a 650 on the verbal portion of the SAT exam will have about a 3.26 GPA at this college.

(c) From the computer output the sample slope is $b_1 = 0.00189$. This means that an additional point on the Verbal SAT exam will give a predicted increase in GPA of 0.00189. (And, likewise, a 100 point increase in verbal SAT will raise predicted GPA by 0.189.)

(d) From the computer output the test statistic for testing $H_0 : \beta_1 = 0$ vs $H_a : \beta_1 \neq 0$ is $t = 6.99$ and the p-value is 0.000. This small p-value gives strong evidence that the verbal SAT score is effective as a predictor of grade point average.

(e) In the output we see that R^2 is 12.5%. This tells us that 12.5% of the variability in grade point averages can be explained by verbal SAT scores.

9.19 (a) Because the variable uses z-scores, we look to see if any of the values for the variable $GMdensity$ are larger than 2 or less than -2. We see that there is only one value outside this range, a point with $GMdensity \approx -2.2$ which is less than -2. This participant has a normalized grey matter score more than two standard deviations below the mean and has about 140 Facebook friends.

(b) On the scatterplot, we have concerns if there is a curved pattern (there isn't) or variability from the line increasing or decreasing (it isn't) or extreme outliers (there aren't any). We do not have any serious concerns about using these data to fit a linear model.

(c) In the output the correlation is $r = 0.436$ and the p-value is 0.005. This is a small p-value so we have strong evidence of a linear relationship between number of Facebook friends and grey matter density.

(d) The least squares line is $\widehat{FBfriends} = 367 + 82.4 \cdot GMdensity$. For a person with a normalized grey matter score of 0, the predicted number of Facebook friends is 367. For a person with $GMDensity$ one standard deviation above the mean, the predicted number of Facebook friends is $367 + 82.4(1) = 449.4$. For a person with grey matter density one standard deviation below the mean, the predicted number of Facebook friends is $367 + 82.4(-1) = 284.6$.

(e) The p-value for a test of the slope is 0.005, exactly matching the p-value for the test of correlation. In fact, if we calculate the t-statistic for testing the correlation using $r = 0.436$ and $n = 40$ we have

$$t = \frac{r\sqrt{n-2}}{\sqrt{1-r^2}} = \frac{0.436\sqrt{40-2}}{\sqrt{1-(0.436)^2}} = 2.99$$

which matches the t-statistic for the slope in the computer output.

(f) We see in the computer output that $R^2 = 19.0\%$. This tell us that 19% of the variability in number of Facebook friends can be explained by the normalized grey matter density in the areas of the brain associated with social perception. Since 19% is not very large, many other factors are involved in explaining the other 81% of the variability in number of Facebook friends.

9.21 (a) For a pH reading of 6.0 we have

$$\widehat{AvgMercury} = 1.53 - 0.152 \cdot pH = 1.53 - 0.152(6) = 0.618$$

The model predicts that fish in lakes with a pH of 6.0 will have an average mercury level of 0.618.

(b) The estimated slope is $b_1 = -0.152$. This means that as pH increases by one unit, predicted average mercury level in fish will go down by 0.152 units.

(c) The test statistic is $t = -5.02$, and the p-value is essentially zero. Since this is a very small p-value we have strong evidence that the pH of a lake is effective as a predictor of mercury levels in fish.

(d) The estimated slope is $b_1 = -0.152$ and the standard error is $SE = 0.03031$. For 95% confidence we use a t-distribution with $53 - 2 = 51$ degrees of freedom to find $t^* = 2.01$. The confidence interval for the slope is

$$\begin{array}{rcl}
b_1 & \pm & t^* \cdot SE \\
-0.152 & \pm & 2.01(0.03031) \\
-0.152 & \pm & 0.0609 \\
-0.2129 & \text{to} & -0.0911
\end{array}$$

Based on these data we are 95% sure that the slope (increase in mercury for a one unit increase in pH) is somewhere between -0.213 and -0.091.

(e) We see that R^2 is 33.1%. This tells us that 33.1% of the variability in average mercury levels in fish can be explained by the pH of the lake water that the fish come from.

9.23 (a) Since 79% gives the percent of variation accounted for, it is a value of R-squared.

(b) Since precipitation is accounting for prevalence of virus, the response variable is prevalence of the virus and the explanatory variable is precipitation.

(c) Since $R^2 = 0.79$, the correlation is $r = \sqrt{0.79} = 0.889$. The correlation might be either 0.889 or -0.889, but we are told that prevalence increased as precipitation increased, so the correlation is positive. We have $r = 0.889$.

9.25 (a) We test: $H_0 : \rho = 0$ vs $H_a : \rho \neq 0$, where ρ represents the population correlation between number of honeybee colonies and year. We use technology to calculate the test statistic and the p-value, or we can calculate them as follows: We have $n = 18$ and $r = -0.41$ so we can calculate the test statistic as

$$t = \frac{r\sqrt{n-2}}{\sqrt{1-r^2}} = \frac{-0.41\sqrt{16}}{\sqrt{1-(-0.41)^2}} = -1.80$$

The proportion beyond -1.80 in the left tail of a t-distribution with 16 df is 0.045. This is a two-tail test, so the p-value is $2(0.045) = 0.09$. At a 5% level, we do not have enough evidence to conclude that the number of bee colonies is linearly related to year.

(b) The percent of the variability refers to the coefficient of determination $R^2 = -0.41^2 = 0.168$. We see that about 16.8% of the variability in number of honeybee colonies can be explained by year.

9.27 (a) The cases are countries of the world.

(b) The scatterplot with the regression line is shown below. We see no obvious curvature and the variability does not change consistently across the plot. The only mild concern we might have is that the negative residuals (points below the line) tend to stretch farther away than the points above the line. So there may be a bit of concern with the symmetry needed for the residuals to be normally distributed.

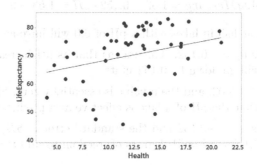

(c) Here is some output for fitting this model.

```
The regression equation is LifeExpectancy = 61.32 + 0.729 Health

Term        Coef  SE Coef  T-Value  P-Value
Constant   61.32     4.70    13.04    0.000
Health     0.729    0.364     2.00    0.051

S = 10.1534    R-Sq =  7.71%    R-Sq(adj) =  5.79%
```

The slope is $b_1 = 0.729$, so we predict that an increase in 1% expenditure on health care would correspond to an increase in life expectancy of about 0.73 years.

(d) We have $df = 48$ so $t^* = 2.011$. We find a 95% confidence interval by taking

$$b_1 \pm t^* \cdot SE = 0.729 \pm 2.011(0.364) = (-0.003, 1.461)$$

We are 95% confident the slope for predicting life expectancy using health expenditure for all countries is between -0.003 and 1.461.

(e) Since our confidence interval (barely!) contains zero we find that the percentage of government expenditure on health may not be an effective predictor of life expectancy at a 5% level. We can also reach this conclusion by considering the p-value ($0.051 > 0.05$) given for testing the slope in the regression output.

(f) The population slope for all countries ($\beta_1 = 0.467$) is not the same as the slope found from this sample ($b_1 = 0.729$), but we see that $\beta_1 = 0.467$ is easily captured within our confidence interval of -0.003 to 1.461.

(g) In the output we see that $R^2 = 7.71\%$. Only 7.71% of the variability in life expectancy in countries is explained by the percentage of the budget spent on health care.

9.29   (a) Here is a scatterplot of life expectancy vs birthrate for all countries.

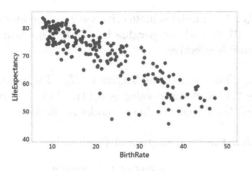

There appears to be a strong negative linear association between birth rate and life expectancy with no obvious curvature and relatively consistent variability.

(b) Yes. We have data on the entire population (all countries) so we can compute $\rho = -0.857$, which is different from 0.

(c) We already have data on (essentially) the entire population so we don't need to use statistical inference to determine where the population correlation might be located.

(d) The slope of the linear regression model is $\beta = -0.7379$, so the predicted life expectancy decreases by about 0.74 years for every one percent increase in birth rate of a country.

(e) We cannot conclude that lowering the birthrate in a country would increase its life expectancy. We cannot make conclusions about causality because this is observational data, not data from a randomized experiment.

Section 9.2 Solutions

9.31 The hypotheses are H_0 : The model is ineffective vs H_a : The model is effective. We see in the ANOVA table that the F-statistic is 21.85 and the p-value is 0.000. This is a small p-value so we reject H_0. We conclude that the linear model is effective.

9.33 The hypotheses are H_0 : The model is ineffective vs H_a : The model is effective. We see in the ANOVA table that the F-statistic is 2.18 and the p-value is 0.141. This is a relatively large p-value so we do not reject H_0. We do not find evidence that the linear model is effective.

9.35 We see in the table that the total degrees of freedom is $n - 1 = 175$, so the sample size is 176. To calculate R^2, we use
$$R^2 = \frac{SSModel}{SSTotal} = \frac{3396.8}{30450.5} = 0.112$$
We see that $R^2 = 11.2\%$.

9.37 We see in the table that the total degrees of freedom is $n - 1 = 343$, so the sample size is 344. To calculate R^2, we use
$$R^2 = \frac{SSModel}{SSTotal} = \frac{10.380}{1640.951} = 0.006$$
We see that $R^2 = 0.6\%$.

9.39 The sum of squares for Model and Error add up to Total sum of squares, so
$$SSE = SSTotal - SSModel = 5820 - 800 = 5020$$

The degrees of freedom for the model is 1, since there is one explanatory variable. The sample size is 40 so the Total df is $40 - 1 = 39$ and the Error df is then $40 - 2 = 38$. We calculate the mean squares by dividing sums of squares by degrees of freedom
$$MSModel = \frac{800}{1} = 800 \qquad \text{and} \qquad MSError = \frac{5020}{38} = 132.1$$

The F-statistic is
$$F = \frac{MSModel}{MSError} = \frac{800}{132.1} = 6.06$$
These values are all shown in the following ANOVA table:

Source	d.f.	SS	MS	F-statistic	p-value
Model	1	800	800	6.06	0.0185
Error	38	5020	132.1		
Total	39	5820			

The p-value is found using the upper-tail of an F-distribution with $df = 1$ for the numerator and $df = 38$ for the denominator. For the F-statistic of 6.06, we see that the p-value is 0.0185.

9.41 The sum of squares for Model and Error add up to Total sum of squares, so

$$SSModel = SSTotal - SSE = 23,693 - 15,571 = 8,122$$

The degrees of freedom for the model is 1, since there is one explanatory variable. The sample size is 500 so the Total df is $500 - 1 = 499$ and the Error df is then $500 - 2 = 498$. We calculate the mean squares by dividing sums of squares by degrees of freedom

$$MSModel = \frac{8122}{1} = 8122 \quad \text{and} \quad MSError = \frac{15571}{498} = 31.267$$

The F-statistic is

$$F = \frac{MSModel}{MSError} = \frac{8122}{31.267} = 259.76$$

These values are all shown in the following ANOVA table:

Source	d.f.	SS	MS	F-statistic	p-value
Model	1	8122	8122	259.76	0.000
Error	498	15571	31.267		
Total	499	23693			

The p-value is found using the upper-tail of an F-distribution with $df = 1$ for the numerator and $df = 498$ for the denominator. We don't really need to look this one up, though. The F-statistic of 259.76 is so large that we can predict that the p-value is essentially zero.

9.43 The hypotheses are H_0 : The model is ineffective vs H_a : The model is effective. We see in the ANOVA table that the F-statistic is 7.44 and the p-value is 0.011. This p-value is significant at a 5% level although not at a 1% level. At a 5% level, we conclude that the linear model to predict calories in cereal using the amount of fiber is effective.

9.45 (a) We see that the predicted grade point average for a student with a verbal SAT score of 550 is given by

$$\widehat{GPA} = 2.03 + 0.00189 VerbalSAT = 2.03 + 0.00189(550) = 3.0695$$

We expect an average GPA of about 3.07 for students who get 550 on the verbal portion of the SAT.

(b) Since the total degrees of freedom is $n - 1 = 344$, the sample size is 345.

(c) To calculate R^2, we use

$$R^2 = \frac{SSModel}{SSTotal} = \frac{6.8029}{54.5788} = 0.125$$

We see that $R^2 = 12.5\%$, which tells us that 12.5% of the variability in grade point averages can be explained by Verbal SAT score.

(d) The hypotheses are H_0 : The model is ineffective vs H_a : The model is effective. We see in the ANOVA table that the F-statistic is 48.84 and the p-value is 0.000. This p-value is very small so we reject H_0. There is evidence that the linear model to predict grade point average using Verbal SAT score is effective.

9.47 (a) We have a small concern about the extreme point in the lower right corner, but otherwise a linear model seems acceptable.

(b) The regression equation is $\widehat{MatingActivity} = 0.480 - 0.323 FemalesHiding$. Predicted mating activity for a group in which the females spend 50% of the time in hiding is

$$\widehat{MatingActivity} = 0.480 - 0.323(0.50) = 0.319$$

(c) For a test of the slope, the hypotheses are $H_0 : \beta_1 = 0$ vs $H_a : \beta_1 \neq 0$. We see in the output that the t-statistic is -2.56 and the p-value is 0.033. At a 5% level we conclude that percent of time females are in hiding is an effective predictor of level of mating activity.

(d) The hypotheses are H_0 : The model is ineffective vs H_a : The model is effective. We see in the ANOVA table that the F-statistic is 6.58 and the p-value is 0.033. At a 5% level we find some evidence that the linear model based on female time spent hiding is effective at predicting mating activity.

(e) The p-values for the t-test and ANOVA are exactly the same. This will always be the case for a simple regression model.

(f) In the output we see that $R^2 = 45.1\%$, so 45.1% of the variability in level of mating activity is explained by the percent of time females spend in hiding.

9.49 (a) To find the standard deviation of the error term, s_ϵ, we use the value of SSE from the ANOVA table in the previous exercise, which is 9979.8, and the sample size of $n = 24$:

$$s_\epsilon = \sqrt{\frac{SSE}{n-2}} = \sqrt{\frac{9979.8}{22}} = 21.3$$

The standard deviation of the error term is 21.3, which we also see in the output as "S = 21.2985."

(b) To find the standard error of the slope, we need to use $s_\epsilon = 21.3$ from part (a). Also, the explanatory variable is $PenMin$ so s_x is the standard deviation of the $PenMin$ values, which is 27.26. Using $n = 24$, we have

$$SE = \frac{s_\epsilon}{s_x \cdot \sqrt{n-1}} = \frac{21.3}{27.26 \cdot \sqrt{23}} = 0.163$$

The standard error of the slope in the model is 0.163, which we also see in the output as "SE Coef" for $PenMin$.

9.51 (a) To find the standard deviation of the error term, s_ϵ, we use the value of SSE from the ANOVA table, which is 27774.1, and the sample size of $n = 30$:

$$s_\epsilon = \sqrt{\frac{SSE}{n-2}} = \sqrt{\frac{27774.1}{28}} = 31.495$$

The standard deviation of the error term is $s_\epsilon = 31.495$.

(b) To find the standard error of the slope, we need to use $s_\epsilon = 31.495$ from part (a). Also, the explanatory variable is $Fiber$ so s_x is the standard deviation of the $Fiber$ values, which is 1.880. Using $n = 30$, we have

$$SE = \frac{s_\epsilon}{s_x \cdot \sqrt{n-1}} = \frac{31.495}{1.880 \cdot \sqrt{29}} = 3.11$$

The standard error of the slope in the model is $SE = 3.11$.

9.53 (a) To test $H_0 : \beta_1 = 0$ vs $H_a : \beta_1 \neq 0$ using a t-test we use the following computer output for this model.

```
Predictor    Coef  SE Coef     T      P
Constant    7.694    1.803  4.27  0.000
TV         0.5955   0.2923  2.04  0.047
```

The test statistic is $t = 2.04$ and the p-value=0.047 which is (barely) significant at a 5% level. We have some evidence that *TV* time is a worthwhile predictor of *Exercise* hours.

(b) For the ANOVA with a single predictor we test $H_0 : \beta_1 = 0$ vs $H_a : \beta_1 \neq 0$ (or the model is ineffective vs the model is effective).

```
Source          DF       SS      MS     F      P
Regression       1   252.27  252.27  4.15  0.047
Residual Error  48  2917.73   60.79
Total           49  3170.00
```

The test statistic is $F = 4.15$ and the p-value is 0.047. Again this is significant at a 5% level and we conclude that *TV* is an effective predictor of *Exercise*.

(c) To compare the correlation between *Exercise* and *TV* we test $H_0 : \rho = 0$ vs $H_a : \rho \neq 0$ where ρ is the correlation between these variables for the population of all college students. Using technology we find the correlation between these two variables for the 50 cases in the sample is $r = 0.282$. We compute the t-statistic as

$$t = \frac{r\sqrt{n-2}}{\sqrt{1-r^2}} = \frac{0.282\sqrt{50-2}}{\sqrt{1-(0.282)^2}} = 2.04$$

Using two-tails and a t-distribution with 48 degrees of freedom, we find p-value= $2(0.0234) = 0.0468$. This is just below a 5% significance level, so we have some evidence of an association between amounts of exercise and TV viewing.

(d) The t-statistics are the same for the tests of slope and correlation (the F-statistic is the square of this value). The p-values are the same for all three tests.

9.55 Here is some computer output for fitting the model to predict *Beds* using *Baths* for the data in **HomesForSaleCA**.

```
Response: Beds
            Estimate Std. Error t value Pr(>|t|)
(Intercept)   1.3671     0.3187   4.289 0.000193 ***
Baths         0.7465     0.1169   6.385 6.53e-07 ***

Residual standard error: 0.6965 on 28 degrees of freedom
Multiple R-squared: 0.5929,    Adjusted R-squared: 0.5783
F-statistic: 40.77 on 1 and 28 DF,  p-value: 6.531e-07
```

(a) The fitted regression equation is $\widehat{Beds} = 1.367 + 0.7465\,Baths$. The predicted number of bedrooms for a house with 3 bathrooms is

$$\widehat{Beds} = 1.367 + 0.7465(3) = 3.6065$$

The average number of bedrooms for homes for sale in California with 3 bathrooms is predicted to be about 3.6 bedrooms.

(b) From the output we see that the t-statistic for testing $H_0 : \beta_1 = 0$ vs $H_a : \beta_1 \neq 0$ is $t = 6.385$ and the p-value that is essentially zero ($6.53x10^{-7}$). This gives strong evidence that the number of bathrooms is an effective predictor of the number of bedrooms.

(c) The F-statistic in the ANOVA table is $F = 40.77$ and the p-value is essentially zero (the same as the t-test). There is strong evidence that this model based on number of bathrooms is effective at predicting the number of bedrooms in houses for sale in California.

(d) In the output we see that $R^2 = 59.3\%$, which means that 59.3% of the variability in the number of bedrooms in houses in California is explained by the number of bathrooms.

9.57 Answers will vary.

Section 9.3 Solutions

9.59 (a) The confidence interval for the mean response is always narrower than the prediction interval for the response, so in this case the confidence interval for the mean response is interval A (94 to 106) and the prediction interval for the response is interval B (75 to 125).

(b) The predicted value is in the center of both intervals, so we can use the average of the endpoints for either interval to find the predicted value is (94+106)/2=100.

9.61 (a) The confidence interval for the mean response is always narrower than the prediction interval for the response, so in this case the confidence interval for the mean response is interval B (19.2 to 20.8) and the prediction interval for the response is interval A (16.8 to 23.2).

(b) The predicted value is in the center of both intervals, so we can use the average of the endpoints for either interval to find the predicted value is (19.2+20.8)/2=20.

9.63 (a) The 95% confidence interval for the mean response is -0.013 to 4.783. We are 95% confident that for mice that eat 10% of calories during the day, the average weight change will be between losing 0.013 grams and gaining 4.783 grams.

(b) The 95% prediction interval for the response is -2.797 to 7.568. We are 95% confident that a mouse that eats 10% of its calories during the day will have a weight change between losing 2.797 grams and gaining 7.568 grams.

9.65 (a) The 95% confidence interval for the mean response is 143.4 to 172.4. We are 95% confident that the average number of calories for all cereals with 16 grams of sugars per cup will be between 143.4 and 172.4 calories per cup.

(b) The 95% prediction interval for the response is 101.5 to 214.3. We are 95% confident that a cereal with 16 grams of sugar will have between 101.5 and 214.3 calories.

9.67 (a) We have $\widehat{Cognition} = 102.3 - 3.34 \cdot Years = 102.3 - 3.34(12) = 62.22$. The predicted cognitive score is 62.2 for a person who has played 12 years of football.

(b) The prediction interval is designed to capture most of the responses while the confidence interval is designed to only capture the mean response. The prediction interval will always be wider, so the 95% confidence interval is given in II, while the 95% prediction interval is given in I.

9.69 (a) Using technology, we see that the 95% prediction interval for life expectancy of a country with 3% expenditure on health care is (41.79, 85.22). We are 95% confident that a country with 3% government expenditure on health care has a life expectancy between 41.8 and 85.2 years.

(b) Using technology, we see that the 95% prediction interval for life expectancy of a country with 10% expenditure on health care is (47.92, 89.29). We are 95% confident that a country with 10% government expenditure on health care has a life expectancy between 47.9 and 89.3 years.

(c) Using technology, we see that the 95% prediction interval for life expectancy of a country with 50% expenditure on health care is (63.33, 132.17). We are 95% confident that a country with 50% government expenditure on health care has a life expectancy between 63.3 and 132.2 years.

(d) We calculate the width of each interval: (a) $85.22 - 41.79 = 43.43$, (b) $89.29 - 47.92 = 41.37$, and (c) $132.17 - 63.33 = 68.84$. We notice that the interval for (a) is slightly larger then (b), and the interval for (c) is much larger then the other two. This is due to the fact that (c) is extrapolating far out from the data used to create the model. In practice we would have much less confidence in an interval for a point that is so far from the data that built the original model – a country with a life expectancy of 132 years would be pretty surprising! The smallest width is (b), where the predictor value (10%) is closest to the mean of the *Health* values (12.3%).

9.71 To calculate 95% confidence and prediction intervals for election margins we use a t-distribution with $12 - 2 = 10$ degrees of freedom to find $t^* = 2.228$. We see from the computer output that the standard deviation of the error is $s_\epsilon = 5.66$ and from the summary statistics for the predictor (*Approval*) that $\overline{x} = 52.92$ and $s_x = 11.04$. We use these values to find the requested intervals.

(a) When the approval rating is 50%, the predicted margin is $\widehat{Margin} = -36.76 + 0.839(50) = 5.19$. We find the 95% confidence interval using

$$\widehat{Approval} \quad \pm \quad t^* s_\epsilon \sqrt{\frac{1}{n} + \frac{(x^* - \overline{x})^2}{(n-1)s_x^2}}$$

$$5.19 \quad \pm \quad 2.228(5.66)\sqrt{\frac{1}{12} + \frac{(50 - 52.92)^2}{11(11.04^2)}}$$

$$5.19 \quad \pm \quad 3.78$$

$$1.41 \quad \text{to} \quad 8.97$$

We are 95% sure that the mean margin of victory for all presidents with 50% approval ratings is between 1.4% and 9.0%.

(b) When the approval rating is 50%, the predicted margin is $\widehat{Margin} = -36.76 + 0.839(50) = 5.19$. We find the 95% prediction interval using

$$\widehat{Approval} \quad \pm \quad t^* s_\epsilon \sqrt{1 + \frac{1}{n} + \frac{(x^* - \overline{x})^2}{(n-1)s_x^2}}$$

$$5.19 \quad \pm \quad 2.228(5.66)\sqrt{1 + \frac{1}{12} + \frac{(50 - 52.92)^2}{11(11.04^2)}}$$

$$5.19 \quad \pm \quad 13.16$$

$$-7.97 \quad \text{to} \quad 18.35$$

We are 95% sure that the margin of victory for a president with a 50% approval rating is between losing by 8.0 points and winning by 18.4 points. We should not have very much confidence in predicting the outcome of this election!

(c) If we have no information about the approval rating, we can find a confidence interval for the mean margin of victory/defeat using an ordinary t-interval for the sample of $n = 12$ margins. From the output we see that $\overline{Margin} = 7.62$ and the standard deviation is 10.72. For 95% confidence with $11 - 1 = 11$ degrees of freedom we have $t^* = 2.201$ and we have

$$\overline{Margin} \quad \pm \quad t^* \frac{s}{\sqrt{n}}$$

$$7.62 \quad \pm \quad 2.201\frac{10.72}{\sqrt{12}}$$

$$7.62 \quad \pm \quad 6.81$$

$$0.81 \quad \text{to} \quad 14.43$$

We are 95% sure that the mean margin of victory for all incumbent presidents, regardless of approval rating, is between 0.8% and 14.4%. Note that this interval is considerably wider than the interval in part (a) that uses the additional information from the approval ratings.

Section 10.1 Solutions

10.1 There are four explanatory variables: $X1$, $X2$, $X3$, and $X4$. The one response variable is Y.

10.3 The predicted response is

$\widehat{Y} = 43.4 - 6.82X1 + 1.70X2 + 1.70X3 + 0.442X4 = 43.4 - 6.82(5) + 1.70(7) + 1.70(5) + 0.442(75) = 62.85$.

The residual is $60 - 62.85 = -2.85$.

10.5 The coefficient of $X1$ is -6.820 and the p-value for testing this coefficient is 0.001.

10.7 Looking at the p-values, we see that variable $X1$ is the only one that is significant at a 1% level.

10.9 Looking at the p-values, we see that $X2$ is least significant.

10.11 We see that $R^2 = 99.8\%$ so 99.8% of the variability in Y is explained by the model.

10.13 The predicted response is

$$\begin{aligned} \widehat{Y} &= -61 + 4.71X1 - 0.25X2 + 6.46X3 + 1.50X4 - 1.32X5 \\ &= -61 + 4.71(15) - 0.25(40) + 6.46(10) + 1.50(50) - 1.32(95) \\ &= 13.85. \end{aligned}$$

The residual is $Y - \widehat{Y} = 20 - 13.85 = 6.15$.

10.15 The coefficient or slope of $X1$ is 4.715 and the p-value for testing this coefficient is 0.053.

10.17 Looking at the p-values, we see that variables $X3$ and $X4$ are significant at a 5% level and variables $X1$, $X2$, and $X5$ are not.

10.19 Looking at the p-values, we see that $X4$ is most significant.

10.21 Yes, the p-value from the ANOVA table is 0.000 so the model is effective.

10.23 (a) The predicted price for a 2500 square foot home with 4 bedrooms and 2.5 baths is

$$\begin{aligned} \widehat{Price} &= -217 + 331SizeSqFt - 135Beds + 200Baths \\ &= -217 + 0.331(2500) - 135(4) + 200(2.5) \\ &= 570.5. \end{aligned}$$

The predicted price for this house is $570,500.

(b) The largest coefficient is 200, the coefficient for the number of bathrooms.

(c) The most significant predictor in this model is $SizeSqFt$ with $t = 4.55$ and p-value $= 0.000$ – even though the size of its coefficient, $\hat{\beta}_1 = 0.331$, is much smaller than the coefficients of the other two predictors.

(d) All three predictors are significant at the 5% level.

(e) If all else stays the same (such as number of bedrooms and number of bathrooms), a house with 1 more square foot in area is predicted to cost 0.331 thousand dollars ($331) more.

(f) The model is effective at predicting price, in the sense that at least one of the explanatory variables are useful, since the p-value is 0.000.

(g) We see that 46.7% of the variability in prices of new homes can be explained by the area in square feet, the number of bedrooms, and the number of bathrooms.

10.25 (a) The last column in the table represents the p-value. We see that both *Internet* and *BirthRate* have p-values less then 0.05, so in this model percentage of the population with internet and the countries birth rate are significant predictors of life expectancy. The p-value for birth rate is much smaller, so this is the most significant predictor in this model.

(b) If we plug these values into our equation we get $\widehat{Life} = 76.24 + 0.131(20) - 0.0003(2.5) + 0.1118(75) - 0.594(30) = 69.42$. So we predict the life expectancy for this country would be 69.42 years.

(c) Since the coefficient of internet is positive ($b_3 = 0.1118$), the predicted life expectancy would increase when internet usage increases.

10.27 (a) For this player, $Years = 9$ and $Concussion = 0$, so the predicted cognition score is $\widehat{Cognition} = 100.6 - 3.07(9) - 2.70(0) = 72.97$. This person's actual cognition score was 74, so the residual is $74 - 72.97 = 1.03$.

(b) For this player, $Years = 7$ and $Concussion = 1$, so the predicted cognition score is $\widehat{Cognition} = 100.6 - 3.07(7) - 2.70(1) = 76.41$. This person's actual cognition score was 42, so the residual is $42 - 76.41 = -34.41$.

(c) The coefficient of *Years* is -3.07. For every additional year playing football, a person's cognitive percentile score on this exam is predicted to go down by 3.07 points, assuming that the person's concussion status doesn't change.

(d) The coefficient of *Concussion* is -2.70. Assuming the same number of years playing football, a person who has had a concussion is predicted to have a cognition score that is 2.70 points lower than a person who has not had a concussion.

(e) Yes, the model based on years of football and concussion status is effective, at a 10% level, for predicting cognitive score, since the ANOVA p-value is 0.05 which is less than 0.10.

(f) We see that the p-value for the test of the *Years* variable is 0.064, while the p-value for the test of the *Concussion* variable is 0.777. Thus, one of the variables (*Years*) is significant at a 10% level, and none are significant at a 5% level.

(g) Since the p-value for *Years* is 0.064, while the p-value for *Concussion* is 0.777, we see that the number of years playing football is more significant than whether or not the player was ever diagnosed with a concussion.

(h) Since df-Total in the ANOVA table is 43, the number of football players included in the analysis is 44.

(i) We see that $R^2 = 13.56\%$, so 13.56% of the variability in cognitive scores for these football players can be explained by the number of years playing football and whether or not the person was ever diagnosed with a concussion.

10.29 The total degrees of freedom, 46, is $n - 1$, so the sample size is 47 horses.

10.31 $H_0 : \beta_1 = \beta_2 = \beta_3 = 0$
H_a : At least one $\beta_i \neq 0$

The p-value in the ANOVA table (0.000) is very small so we have strong evidence to reject H_0 and conclude that at least one of the predictors in the model is useful for explaining horse prices.

10.33 Here is some computer output for this model:

```
Coefficients:
            Estimate Std. Error t value Pr(>|t|)
(Intercept)   -83.49     161.52  -0.517 0.609593
Size          483.89     124.21   3.896 0.000613 ***
Beds         -254.86      84.37  -3.021 0.005594 **
Baths         228.92     104.32   2.194 0.037337 *
```

(a) We see that all three of our predictors are significant at a 5% level, and *Size* is the most significant with a p-value of only 0.0006.

(b) We interpret the coefficient of *Beds* (-254.86) by saying that if the size of the house and the number of bathrooms were to stay the same, an increase of one bedroom in a house will lower the predicted cost by \$254,860. This might not make sense, as we expect one more bedroom to increase the cost not decrease. However, more rooms in the same space means smaller rooms and smaller rooms might decrease the value of a house. Remember that we need to interpret coefficients in the context of the other variables! For *Baths* we interpret the coefficient of 228.92 by saying that if the size of the house and the number of bedrooms were to stay constant, an increase in one bathroom will raise the predicted cost of a house by \$228,920. This makes much more sense.

(c) We predict the price would be $-83.49 + 483.89(1.5) - 254.86(3) + 228.92(2) = 335.605$ or \$335,605.

10.35 (a) Here is some output for fitting $WinPct = \beta_0 + \beta_1 PtsFor + \beta_2 PtsAgainst + \epsilon$.

```
Term           Coef  SE Coef  T-Value  P-Value
Constant      0.492    0.245     2.01    0.055
PtsFor      0.03223  0.00173    18.62    0.000
PtsAgainst -0.03215  0.00176   -18.31    0.000


        S    R-sq  R-sq(adj)
0.0352418  95.97%     95.67%


Analysis of Variance
Source      DF       SS        MS  F-Value  P-Value
Regression   2  0.79792  0.398962   321.23    0.000
Error       27  0.03353  0.001242
Total       29  0.83146
```

The estimated prediction equation is $\widehat{WinPct} = 0.492 + 0.03223 \cdot PtsFor - 0.03215 \cdot PtsAgainst$

(b) The predicted winning percentage for the Golden State Warriors with $PtsFor = 114.9$ and $PtsAgainst = 104.1$ is

$$\widehat{WinPct} = 0.492 + 0.03223 \cdot 114.9 - 0.03215 \cdot 104.1 = 0.848$$

The residual is $WinPct - \widehat{WinPct} = 0.890 - 0.848 = 0.042$ which is a bit better than the residual based on $PtsFor$ alone and much better than the residual based on a model with just $PtsAgainst$.

(c) The t-statistics for both coefficients (18.62 and -18.31) are extremely large in magnitude (and very similar in size) and both p-values are essentially zero, so we have strong evidence that both offensive ($PtsFor$) and defensive ($PtsAgainst$) abilities are important for predicting team success in the NBA (as measured by $WinPct$).

(d) Using just $PtsFor$ as a predictor for $WinPct$ we find $R^2 = 45.87\%$ and $s_\epsilon = 0.1268$. Using just $PtsAgainst$ as a predictor for $WinPct$ we find $R^2 = 44.16\%$ and $s_\epsilon = 0.1288$. Both single predictor models have ANOVA p-value≈ 0.000 so are very effective (and about equally so). But together in a two-predictor model we see $R^2 = 95.97\%$ (much higher than the single predictors) and $s_\epsilon = 0.0352$ (much lower than the single predictors). Thus the model based on *both* predictors is much more effective at predicting winning percentage in the NBA.

10.37 (a) Using technology we find R^2 for each model.

Predictor(s)	R^2
$PtsFor$	45.87%
$PtsAgainst$	44.16%
$PtsFor, PtsAgainst$	95.97%

The combination of the two predictors works much better than either alone – even better than the sum of the two parts.

(b) Here is some output for fitting $\widehat{WinPct} = \beta_0 + \beta_1 Diff + \epsilon$.

```
The regression equation is
WinPct = 0.50011 + 0.03219 Diff

Term         Coef   SE Coef   T-Value   P-Value
Constant   0.50011  0.00632     79.15     0.000
Diff       0.03219  0.00125     25.81     0.000

        S    R-sq   R-sq(adj)
0.0346074  95.97%     95.82%

Analysis of Variance
Source      DF       SS       MS    F-Value   P-Value
Regression   1  0.79792  0.797922    666.23     0.000
Error       28  0.03353  0.001198
Total       29  0.83146
```

The prediction equation is $\widehat{WinPct} = 0.500 + 0.0322 Diff$. This is a very effective model (t=25.81, p-value ≈ 0 for the t-test for slope) and the $R^2 = 95.97\%$ matches that of the two-predictor model (and is much better than either predictor individually). The scatterplot with the regression line shows a strong linear pattern with small amounts of random scatter above and below the line. Note also that the intercept is essentially 0.500, showing that a team that scores as much as it allows should win about half of its games.

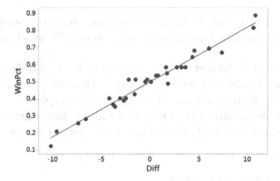

Section 10.2 Solutions

10.39 We see that the residuals are all positive at the ends and negative in the middle. This matches the curved pattern we see in graph (c).

10.41 We see that there is a large outlier in the scatterplot. Since the outlier is above the line, the residual is positive. Also, the outlier matches a predicted value of about 29, which matches what we see in graph (a).

10.43 The conditions all seem to be met. There appears to be a mild linear trend in the data (although it is not very strong), with no reason to question the equal variability condition. We see no strong skewness or outliers in the histogram of the residuals. Finally, the scatterplot of residuals against predicted values appears to show the residuals equally spread out above and below a line at 0, indicating no obvious trend or curvature, and the spread of the points above/below the line stays reasonably consistent going across the graph.

10.45   (a) The arrow is shown on the figure.

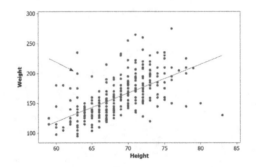

(b) The predicted weight is $\widehat{Weight} = -170 + 4.82 Height = -170 + 4.82(63) = 133.66$ lbs. The residual is $200 - 133.66 = 66.34$.

(c) Arrows are shown on both the figures.

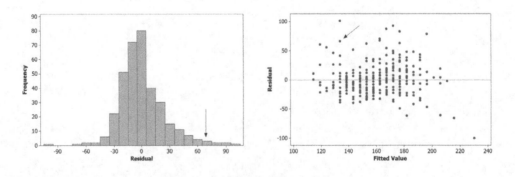

(d) Looking at the three graphs, the conditions appear to all be met.

10.47 The three relevant plots, a scatterplot with least squares line, a histogram of the residuals, and a residuals versus fitted values plot are shown below.

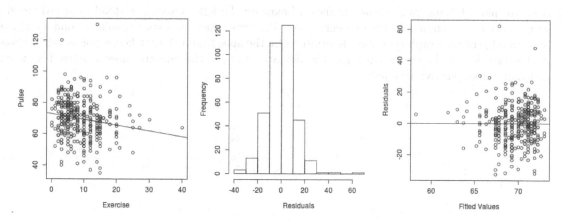

The conditions appear to be reasonably met. There are a few outliers which may be cause for concern (a couple outliers with high residuals, and an outlier with a high exercise amount), but these outliers are not too extreme, and mostly the conditions appear to be met. The linear trend does not appear to be particularly strong, but there is no evidence of a non-linear trend. The histogram is roughly normal, and variability appears to be constant.

10.49 The three relevant plots, a scatterplot with least squares line, a histogram of the residuals, and a residuals versus fitted values plot are shown below.

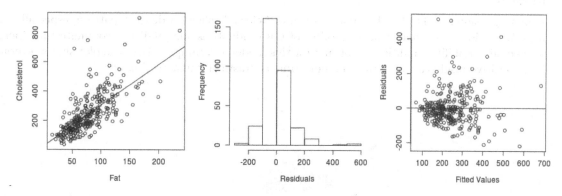

The conditions appear to be reasonably well met. There appears to be a linear trend in the scatterplot with reasonably equal variability around the regression line. The histogram of residuals appears to have a couple of mild outliers, but nothing too extreme. The scatterplot of residuals vs fits seems to have no obvious trend or curvature and relatively equal variability. The conditions aren't perfectly met but seem reasonably close.

10.51   (a) Here is some output for fitting this model

```
              Estimate Std. Error t value Pr(>|t|)
(Intercept)    7.12034    0.90659   7.854  2.5e-14 ***
Distance       1.21115    0.03977  30.454  < 2e-16 ***
```

The prediction equation is $\widehat{Time} = 7.12 + 1.211 \cdot Distance$.

(b) For *Distance* = 20 the expected commute time for the model is $\widehat{Time} = 7.12 + 1.211 \cdot 20 = 31.34$ minutes.

(c) A scatterplot of *Time* vs *Distance* is shown below on the left. There is a steady upward trend with longer commutes tending to take longer times. However, there is a very regular boundary along the bottom edge of the graph (possibly determined by the speed limit?) that keeps points fairly close and in the direction of the regression line. The deviations above the line are more scattered with several unusually large positive residuals.

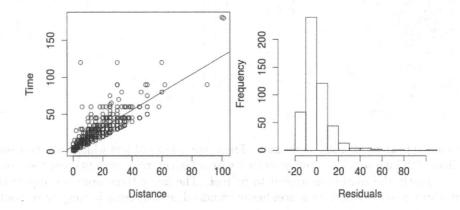

(d) A histogram of the residuals is shown to the right above. It shows a clear skew to the right with a long tail and several large outliers in that direction. The normality condition is not appropriate for these residuals.

(e) A plot of residual vs fits for this model is shown below. It shows a distinct pattern, especially for the negative residuals, of increasing variability of the residuals as the predicted commute time increases. The equal variability condition is not met for this model and sample. Also, the plot shows several very large positive residuals where the commute time is drastically underestimated.

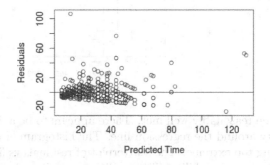

10.53 (a) We are 95% sure that the mean commute time for all Atlantans with a 20 mile commute is between 30.3 and 32.4 minutes.

(b) The lower bound of the prediction interval implies a 20 mile commute in 7.235 minutes, which is an average speed of $(20/7.235) \cdot 60 \approx 166$ miles per hour! The prediction interval assumes normally distributed, hence symmetric, errors. A symmetric interval around the expected commute time needs to be very wide to capture most of the positively skewed times, which makes the lower bound unreasonably small.

10.55 A histogram of the residuals and plot of residuals versus fitted values are shown below. The histogram is nicely symmetric with no extreme outliers. The normality condition is reasonable. The residuals are randomly scattered in a band on either side of the zero line in the residual vs fits plot, with no obvious trend or curvature. We see no concerns with the constant variability condition.

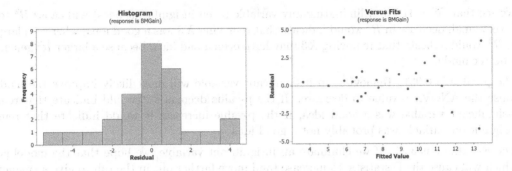

10.57 A histogram of the residuals and plot of residuals versus fitted values are shown below. The histogram appears to have a large potential outlier which raises minor concerns. The residuals appear to have somewhat greater variability for large fits than for small fits, so we have some concern about the constant variability condition. Also, there is some curvature downward and then an upward trend.

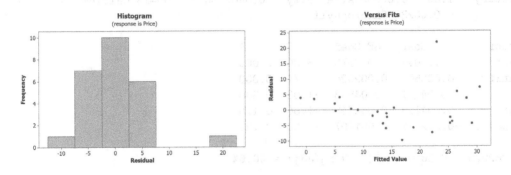

Section 10.3 Solutions

10.59 (a) We should try eliminating the variable $X3$ since it has the largest p-value and hence is the most insignificant in this model.

(b) We see that $R^2 = 41.7\%$. Eliminating any variable (even insignificant ones) will cause R^2 to decrease. A very small decrease in R^2 would indicate that removing $X3$ was a good idea, whereas a large decrease in R^2 would indicate that removing $X3$ may have been a bad idea, because a larger R^2 generally means a better model.

(c) The p-value is 0.031. Removing an insignificant variable will most likely improve the model, and so cause the ANOVA p-value to decrease. If the p-value decreases it would indicate that removing the insignificant variable was a good idea, if the p-value increases it would indicate that removing the insignificant variable was probably not a good idea.

(d) The F-statistic is 3.81. If we eliminate an insignificant variable, we hope that the model gets better, which will cause the F-statistic to increase (and move farther out in the tail to give a smaller p-value.) An increase in the F-statistic would indicate that removing $X3$ was a good idea.

10.61 Here is the output for fitting all four predictors:

```
The regression equation is
Avg_Mercury = 1.00 - 0.00550 Alkalinity - 0.0467 pH + 0.00413 Calcium
              - 0.00236 Chlorophyll


Predictor         Coef   SE Coef      T      P
Constant        1.0044    0.2576   3.90  0.000
Alkalinity   -0.005503  0.002028  -2.71  0.009
pH            -0.04671   0.04533  -1.03  0.308
Calcium       0.004129  0.002648   1.56  0.125
Chlorophyll  -0.002361  0.001497  -1.58  0.121


S = 0.262879   R-Sq = 45.2%   R-Sq(adj) = 40.6%

Analysis of Variance
Source          DF       SS       MS     F      P
Regression       4  2.73081  0.68270  9.88  0.000
Residual Error  48  3.31707  0.06911
Total           52  6.04788
```

The ANOVA p-value is already very low for this model, so it is clear the model is effective. However, some of the variables are not significant. The least significant variable is pH so we see what happens when we eliminate that one.

```
The regression equation is
Avg_Mercury = 0.745 - 0.00649 Alkalinity + 0.00433 Calcium - 0.00303 Chlorophyll


Predictor         Coef   SE Coef      T      P
Constant       0.74458   0.05240  14.21  0.000
Alkalinity   -0.006487  0.001790  -3.62  0.001
Calcium       0.004333  0.002642   1.64  0.107
```

```
Chlorophyll  -0.003035  0.001348  -2.25  0.029
```

```
S = 0.263045   R-Sq = 43.9%   R-Sq(adj) = 40.5%
```

Analysis of Variance

Source	DF	SS	MS	F	P
Regression	3	2.65743	0.88581	12.80	0.000
Residual Error	49	3.39045	0.06919		
Total	52	6.04788			

The ANOVA p-value is still zero, but we can see that the model is improved since the F-statistic went up. The value of R^2 went down, but only by a bit. The p-values of the other three variables all improved. The only one that is not significant at a 5% level is *Calcium* so we see what happens if we eliminate that variable.

The regression equation is
Avg_Mercury = 0.752 - 0.00415 Alkalinity - 0.00298 Chlorophyll

Predictor	Coef	SE Coef	T	P
Constant	0.75194	0.05308	14.17	0.000
Alkalinity	-0.004154	0.001105	-3.76	0.000
Chlorophyll	-0.002979	0.001370	-2.17	0.034

```
S = 0.267451   R-Sq = 40.9%   R-Sq(adj) = 38.5%
```

Analysis of Variance

Source	DF	SS	MS	F	P
Regression	2	2.4714	1.2357	17.27	0.000
Residual Error	50	3.5765	0.0715		
Total	52	6.0479			

The F-statistic again went up, and again R^2 went down from 43.9% to 40.9%. Both remaining variables are significant at a 5% level. However, s_ϵ increased from 0.263 to 0.267, and Adjusted R^2 decreased. Either this model or the model before eliminating *Calcium* could be justified as the best model.

10.63 (a) Here are the correlations between *Time* and each predictor, along with p-values for testing if each correlation differs from zero.

	Runs	Margin	Hits	Errors	Pitchers	Walks
Correlation with *Time*	0.504	-0.116	0.349	-0.040	0.721	0.565
p-value	0.005	0.541	0.059	0.833	0.000	0.001

Pitchers appear to be the strongest predictor of *Time* as a single predictor with the largest correlation and smallest p-value. *Walks* and *Runs* are also strong single predictors of *Time* with very small p-values. The correlation between *Hits* and *Time* is not quite significant at a 5% level. *Margin* and *Error* have the weakest correlations and do not appear to be useful predictors of *Time* on their own.

  (b) If we try the three strongest individual predictors, *Pitchers*, *Walks* and *Runs*, we get the following output.

Predictor	Coef	SE Coef	T	P
Constant	120.083	9.629	12.47	0.000
Runs	0.7016	0.7086	0.99	0.331

```
Walks        2.1966    0.9813    2.24    0.034
Pitchers     5.255     1.449     3.63    0.001

S = 13.3518    R-Sq = 62.1%    R-Sq(adj) = 57.7%
```

Pitchers (p-value=0.001) and *Walks* (p-value=0.034) both look effective in this model, but we might drop *Runs* (p-value=0.331). Doing so we get the output below.

```
Predictor      Coef   SE Coef      T      P
Constant    120.550     9.614  12.54  0.000
Walks         2.3632    0.9664   2.45  0.021
Pitchers      5.851     1.317    4.44  0.000

S = 13.3470    R-Sq = 60.7%    R-Sq(adj) = 57.8%
```

Both predictors are significant and, while the R^2 has dropped by 1.4%, the estimate of the standard deviation of the error, both adjusted R^2 and $s_\epsilon = 13.347$ are slightly better here than the three predictor model.

We tried each of the other predictors together with *Pitchers* and *Walks* and none of them showed a significant p-value for their coefficient. So we will use the two predictor model, $\widehat{Time} = 120.55 + 2.3632 \cdot Walks + 5.851 \cdot Pitchers$.

10.65 (a) We regress *LifeExpectancy* on *Cell* and obtain the following output:

```
Coefficients:
             Estimate Std. Error t value Pr(>|t|)
(Intercept)  59.9185      1.4537   41.22  < 2e-16 ***
Cell          0.1030      0.0128    8.03  9.9e-14 ***
---
Residual standard error: 7.93 on 191 degrees of freedom
  (24 observations deleted due to missingness)
Multiple R-squared:  0.252,    Adjusted R-squared:  0.248
```

The p-value of 9.9×10^{-14} (essentially zero) indicates that *Cell* is a very significant predictor of *LifeExpectancy*.

(b) GDP is associated with both number of mobile subscriptions ($r = 0.448$) and with life expectancy ($r = 0.587$), so is a potential confounding variable.

(c) We regress *LifeExpectancy* on both *Cell* and *GDP* and obtain the following output:

```
Coefficients:
              Estimate Std. Error t value Pr(>|t|)
(Intercept) 60.1467995  1.3883599   43.32  < 2e-16 ***
Cell         0.0696329  0.0133304    5.22  4.9e-07 ***
GDP          0.0001952  0.0000282    6.94  7.4e-11 ***
```

```
---
Residual standard error: 6.99 on 176 degrees of freedom
  (38 observations deleted due to missingness)
Multiple R-squared:  0.434,     Adjusted R-squared:  0.428
F-statistic: 67.6 on 2 and 176 DF,  p-value: <2e-16
```

The p-value of 4.97×10^{-7} for *Cell* is very small, indicating that even after accounting for *GDP*, *Cell* is a very significant predictor of *LifeExpectancy*.

10.67 (a) The predicted commute time for the steel bike is

$$\widehat{Minutes} = 108.342 - 0.553(1) = 107.789 \text{ minutes.}$$

(b) The predicted commute time for the carbon bike is

$$\widehat{Minutes} = 108.342 - 0.553(0) = 108.342 \text{ minutes.}$$

(c) The p-value of 0.711 for testing the coefficient of *BikeSteel* indicates no significant difference in average commute time between the carbon bike and the steel bike.

10.69 (a) For commutes of the same distance, the predicted commute time is 3.571 minutes longer for the steel bike than for the carbon bike.

(b) For commutes on the same bike, the predicted commute time increases by 10.5 minutes for every mile ridden.

(c) The predicted commute time for a 27 mile commute on the steel bike is

$$\widehat{Minutes} = -176.62 + 3.571(1) + 10.41(27) = 108.021 \text{ minutes.}$$

The predicted commute time for a 27 mile commute on the carbon bike is

$$\widehat{Minutes} = -176.62 + 3.571(0) + 10.41(27) = 104.45 \text{ minutes.}$$

Unit D: Essential Synthesis Solutions

D.1 This is a chi-square goodness-of-fit test. The null hypothesis is that the bills are equally distributed among the three servers while the alternative hypothesis is that they are not equally spread out. In symbols the hypotheses are

$$H_0 : \quad p_A = p_B = p_C = 1/3$$
$$H_a : \quad \text{Some } p_i \neq 1/3$$

The expected count for each cell is $n \cdot p_i = 157 \cdot (1/3) = 52.33$. We compute the chi-square statistic:

$$\chi^2 = \frac{(60 - 52.33)^2}{52.33} + \frac{(65 - 52.33)^2}{52.33} + \frac{(32 - 52.33)^2}{52.33} = 1.124 + 3.068 + 7.898 = 12.09$$

The upper-tail p-value from a chi-square distribution with $df = 2$ is 0.002. This is a small p-value so we reject H_0 and find evidence that the bills are not equally distributed between the three servers. Server C appears to have substantially fewer bills than expected if they were equally distributed, and the result is significant enough to generalize (assuming the sample data are representative of all bills).

D.3 This is a chi-square test on a two-way table. The null hypothesis is that use of a credit card does not differ based on the server and the alternative hypothesis is that there is an association between card use and server.

The expected count for the (Cash, Server A) cell is

$$(60 \cdot 106)/157 = 40.51$$

For the (Cash, Server A) cell, we then compute the contribution to the chi-square statistic as

$$(39 - 40.51)^2/40.51 = 0.056$$

Finding the other expected counts and contributions similarly (or using technology) produces a table with the observed counts in each cell, expected counts below them, and the contributions to the chi-square statistic below that, as in the computer output below.

```
                A        B        C
Cash           39       50       17
            40.51    43.89    21.61
           0.0563   0.8520   0.9816

Card           21       15       15
            19.49    21.11    10.39
           0.1169   1.7708   2.0401


Cell Contents:       Count
                     Expected count
                     Contribution to Chi-square


Pearson Chi-Square = 5.818, DF = 2, P-Value = 0.055
```

Adding up all the contributions to the chi-square statistic, we obtain $\chi^2 = 5.818$ (also seen in the computer output). Using the upper tail of a chi-square distribution with $df = 2$, we obtain a p-value of 0.055. This is a borderline p-value, but, at a 5% level, does not provide enough evidence of an association between server and the use of cash or credit.

D.5 This is an analysis of variance test for a difference in means. The sample sizes of the three groups (servers) are all greater than 30, so the normality condition is met and we see from the standard deviations that the condition of relatively equal variability is also met. We proceed with the ANOVA test.

The null hypothesis is that the means are all the same (no association between tip percent and server) and the alternative hypothesis is that there is a difference in mean tip percent between the servers. In symbols, where μ represents the mean tip percentage, the hypotheses are

$$H_0: \quad \mu_A = \mu_B = \mu_C$$
$$H_a: \quad \text{Some } \mu_i \neq \mu_j$$

Using technology with the data in **RestaurantTips** we obtain the analysis of variance table shown below.

One-way ANOVA: PctTip versus Server

Source	DF	SS	MS	F	P
Server	2	83.1	41.6	2.19	0.115
Error	154	2917.9	18.9		
Total	156	3001.1			

We see that the F-statistic is 2.19 and the p-value for an F-distribution with 2 and 154 degrees of freedom is 0.115. This is not a small p-value so we do not reject H_0. We do not find convincing evidence of a difference in mean percentage tip between the three servers.

D.7 (a) There appears to be a positive association in the data with larger parties tending to give larger tips. There is one outlier: The (generous) party of one person who left a $15 tip.

(b) We are testing $H_0 : \rho = 0$ vs $H_a : \rho > 0$ where ρ is the correlation between number of guests and size of the tip. As always, the null hypothesis is that there is no relationship while the alternative hypothesis is that there is a relationship between the two variables. The test statistic is

$$t = \frac{r \cdot \sqrt{n-2}}{\sqrt{1-r^2}} = \frac{0.504 \cdot \sqrt{155}}{\sqrt{1-(0.504^2)}} = 7.265$$

This is a one-tailed test, so the p-value is the area above 7.265 in a t-distribution with $df = n - 2 = 155$. We see that the p-value is essentially zero. There is strong evidence of a significant positive linear relationship between these two variables. The tip does tend to be larger if there are more guests.

(c) No, we cannot assume causation since these data do not come from an experiment. An obvious confounding variable is the size of the bill, since the bill tends to be higher for more guests and higher bills generally tend to correspond to higher tips.

D.9 (a) We see that $R^2 = 83.7\%$, which tells us that 83.7% of the variability in the amount of the tip is explained by the size of the bill.

(b) From the output the F-statistic is 797.87 and the p-value is 0.000. There is very strong evidence that this regression line $\widehat{Tip} = -0.292 + 0.182 \cdot Bill$ is effective at predicting the size of the tip.

Unit D: Review Exercise Solutions

D.11 The area above $t = 1.36$ gives a p-value of 0.0970, which is not significant at a 5% level.

D.13 For 5 groups we have $5 - 1 = 4$ degrees of freedom for the chi-square goodness-of-fit statistic. The area above $\chi^2 = 4.18$ gives a p-value of 0.382, which is not significant at a 5% level.

D.15 For a difference in means ANOVA with $k = 6$ groups and and an overall sample size of $n = 100$ we use an F-distribution with $k - 1 = 6 - 1 = 5$ numerator df and $n - k = 100 - 6 = 94$ denominator df. The area above $F = 2.51$ for this distribution gives a p-value of 0.035, which is significant at a 5% level.

D.17 For a multiple regression model with $k = 3$ predictors we have $n - k - 1 = 26 - 3 - 1 = 22$ degrees of freedom for the t-distribution to test an individual coefficient. Accounting for two tails and the area beyond $t = 1.83$ gives a p-value of $2(0.0404) = 0.0808$, which is not significant at a 5% level.

D.19 White blood cell count is a quantitative variable so this is a test for a difference in means between the three groups, which is analysis of variance for difference in means.

D.21 Both of the relevant variables are quantitative, and we can use a test for correlation, a test for slope, or ANOVA for regression.

D.23 The data are counts from the sample within different racial categories. To compare these with the known racial proportions for the entire city, we use a chi-square goodness-of-fit test with the census proportions as the null hypothesis.

D.25 They are using many quantitative variables to develop a multiple regression model. To test its effectiveness, we use ANOVA for regression.

D.27 Whether or not a case gets settled out of court is categorical, and the county is the other categorical variable, so this is a test between two categorical variables. A chi-square test for association is most appropriate.

D.29 The relevant hypotheses are $H_0 : p_s = p_r = p_g = p_p = 0.25$ vs H_a : Some $p_i \neq 0.25$, where the p_i's are the proportions in the Since we are only interested in cases where one performed higher on the math and verbal sections we ignore the students who scored the same on each. The relevant hypotheses are $H_0 : p_m = p_v = 0.5$ vs H_a : Some $p_i \neq 0.5$, where the p_m and p_v are the proportions with higher Math or Verbal SAT scores, respectively.

The total number of students (ignoring the ties) is $205 + 150 = 355$ so the expected count in each cell, assuming equally likely, is $355(0.5) = 177.5$. We compute a chi-square test statistic as

$$\chi^2 = (205 - 177.5)^2/177.5 + (150 - 177.5)^2/177.5 = 8.52$$

Using a the upper tail of chi-square distribution with 1 degree of freedom yields a p-value of 0.0035. At a 5% significance level, we conclude that students are not equally likely to have higher Math or Verbal scores. From the data, we see that students from this population are more likely to have a higher Math score. (Note that since there are only two categories after we eliminate the ties, we could have also done this problem as a z-test for a single proportion with $H_0 : p = 0.5$.)

D.31 (a) To see if a chi-square distribution is appropriate, we find the expected count in each cell by multiplying the total for the row by the total for the column and dividing by the overall sample size ($n = 85$). These expected counts are summarized in the table below and we see that the smallest expected count of 5.2 (Rain, Fall) is (barely) greater than five, so a chi-square distribution is reasonable.

	Spring	Summer	Fall	Winter	Total
Rain	5.4	5.7	5.2	5.7	22
No Rain	15.6	16.3	14.8	16.3	63
Total	21	22	20	22	85

(b) The null hypothesis is that distribution of rain/no rain days in San Diego does not depend on the season and the alternative is that the rain/no rain distribution is related to the season. We have already calculated the expected counts in part (a) so we proceed to compute the chi-square test statistic by summing $(observed - expected)^2/expected$ for each cell

$$\chi^2 = \frac{(5 - 5.4)^2}{5.4} + \frac{(0 - 5.7)^2}{5.7} + \frac{(6 - 5.2)^2}{5.2} + \cdots + \frac{(11 - 16.3)^2}{16.3} = 14.6$$

Comparing $\chi^2 = 14.6$ to the upper tail of a chi-square distribution with 3 degrees of freedom yields a p-value of 0.002.

(c) This is a small p-value, so we reject the null hypothesis, indicating that there is a difference in the proportion of rainy days among the four seasons, and it appears the rainy season (with almost twice as many rainy days as expected if there were no difference) is the winter.

D.33 (a) The null hypothesis is that all the state average home prices are equal and the alternative is that at least two states have different means.

$H_0 : \mu_{CA} = \mu_{NY} = \mu_{NJ} = \mu_{PA}$
$H_a :$ Some $\mu_1 \neq \mu_j$

(b) We are comparing $k = 4$ states, so the numerator degrees of freedom is $k - 1 = 3$.

(c) The overall sample size is 120 homes, so the denominator degrees of freedom is $n - k = 116$.

(d) The sum of squares for error will tend to be much greater than the sum of squares for groups because we will be dividing the sum of squares for error by 116 to standardize it, while we will only be dividing the sum of squares groups by 3. However, without looking at the data, we cannot tell for sure and we have even less knowledge about how the *mean squares* might compare.

D.35 The null hypothesis is that the mean fiber amounts are all the same and the alternative hypothesis is that at least two of the companies have different mean amounts of fiber.

$H_0 : \mu_{GM} = \mu_K = \mu_Q$
$H_a :$ Some $\mu_1 \neq \mu_j$

Since there are 3 groups (the three companies), the degrees of freedom for groups is 2. Since the sample size is 30, the total degrees of freedom is 29. The error degrees of freedom is $30 - 3 = 27$. We subtract to find the error sum of squares: $SSError = SSTotal - SSGroups = 102.47 - 4.96 = 97.51$. Filling in the rest of the ANOVA table, we have:

```
Source    DF      SS     MS     F      P
Company   2     4.96   2.48   0.69   0.512
Error     27   97.51   3.61
Total     29  102.47
```

The p-value of 0.512 is found using the upper tail (beyond $F = 0.69$) for an F-distribution with 2 numerator df and 27 denominator df. The p-value is very large so this sample does not provide evidence that the mean number of grams of fiber differs between the three companies.

D.37 (a) For the red ink sample, the mean is 4.4 and the sample size is 19. We have $MSE = 0.84$ and, for 95% confidence from a t-distribution with $df = n - k = 71 - 3 = 68$, we have $t^* = 2.00$. The confidence interval is

$$\overline{x}_i \quad \pm \quad t^* \frac{\sqrt{MSE}}{\sqrt{n_i}}$$

$$4.4 \quad \pm \quad 2.00 \frac{\sqrt{0.84}}{\sqrt{19}}$$

$$4.4 \quad \pm \quad 0.42$$

$$3.98 \quad \text{to} \quad 4.82$$

We are 95% confident that the mean number of anagrams solved by people getting the puzzles in red ink is between 3.98 and 4.82.

(b) Using the same t^* and MSE from part (a), we have

$$(\overline{x}_i - \overline{x}_j) \quad \pm \quad t^* \sqrt{MSE \left(\frac{1}{n_i} + \frac{1}{n_j} \right)}$$

$$(5.7 - 4.4) \quad \pm \quad 2.00 \sqrt{0.84 \left(\frac{1}{27} + \frac{1}{19} \right)}$$

$$1.3 \quad \pm \quad 0.55$$

$$0.75 \quad \text{to} \quad 1.85$$

We are 95% confident that people can solve, on average, between 0.75 and 1.85 more anagrams when green ink is used than when red ink is used. Since zero is not in this interval, there is a significant difference between the mean number of anagrams solved in the two conditions.

(c) We are testing $H_0 : \mu_R = \mu_B$ vs $H_a : \mu_R \neq \mu_B$ where μ_R and μ_B represent the mean number of anagrams people can solve if the ink used is red or black, respectively. The test statistic is

$$t = \frac{\overline{x}_R - \overline{x}_B}{\sqrt{MSE \left(\frac{1}{n_R} + \frac{1}{n_B} \right)}} = \frac{4.4 - 5.9}{\sqrt{0.84 \left(\frac{1}{19} + \frac{1}{25} \right)}} = -5.38$$

We use a t-distribution with $df = 68$ to find the p-value. The t-statistic is quite large in magnitude so, even after multiplying by 2 for the two-tailed test, the p-value is essentially zero. There is very strong evidence of a difference in the mean number of anagrams people can solve based on whether the ink is red or black. We see that people solve fewer anagrams with red ink.

D.39 The hypotheses are $H_0 : \rho = 0$ vs $H_a : \rho > 0$, where ρ is the correlation between standardized cognition score and GPA for all students. The t-statistic is:

$$t = \frac{r\sqrt{n-2}}{\sqrt{1-r^2}} = \frac{0.267\sqrt{251}}{\sqrt{1-(0.267^2)}} = 4.39$$

Using a t-distribution with $n - 2 = 253 - 2 = 251$ degrees of freedom, we find a p-value of essentially zero. We find strong evidence of a positive association between $CognitionZscore$ and GPA for students at this college.

D.41 (a) The slope is $b_1 = 0.0831$. If the depression score goes up by 1, the predicted number of classes missed goes up by 0.0831.

(b) The t-statistic is 2.47 and the p-value is 0.014. At a 5% level, we find that the depression score is an effective predictor of the number of classes missed.

(c) We see that $R^2 = 2.4\%$. Only 2.4% of the variability in number of classes missed can be explained by the depression score. Clearly, many other variables are also involved in explaining the number of classes a student will miss in a semester.

(d) The F-statistic is 6.09 and the p-value is 0.014. At a 5% level, we find that this model based on the depression score has some value for predicting the number of classes missed.

D.43 There are several problems with the regression conditions, the most serious of which is the number of large outliers; points well above the line in the scatterplot with regression line. These also contribute to the right skew in the histogram of residuals, violating the normality condition. The residuals vs fits plot doesn't show roughly equal bands on either side of the zero mean, rather we again see the several large positive residuals that aren't balanced with similar sized negative residuals below the line. There is no clear curvature in the data, but the residual vs fits plot shows an interesting pattern as the most extreme negative residuals decrease in regular fashion – not a random scatter. We should be hesitant to use inference based on a linear model for these data (including the earlier exercise for these variables).

D.45 (a) The 95% confidence interval for the mean response is 692.2 to 759.8. We are 95% confident that the mean number of points for all players who make 100 free throws in a season is between 692.2 and 759.8.

(b) The 95% prediction interval for the response is 339.3 to 1112.7. We are 95% confident that a player who makes 100 free throws in a season will have between 339.3 and 1112.7 points for the season.

D.47 (a) The coefficient of *Gender* is -0.0971. Since males are coded 1, this means that, all else being equal, a male is predicted to have a GPA that is 0.0971 less than a female. The coefficient of *ClassYear* is -0.0558. All else being equal, as students move up one class year, their GPA is predicted to go down by 0.0558. The coefficient of *ClassesMissed* is -0.0146. All else being equal, for every additional class missed, the predicted GPA of students goes down by 0.0146.

(b) The p-value from the ANOVA test is 0.000 so this model is effective at predicting GPA.

(c) We see that $R^2 = 18.4\%$ so 18.4% of the variability in grade point averages can be explained by the model using these six explanatory variables.

(d) We see from the p-values for the individual slopes that *CognitionZscore* is the most significant variable in the model, with a p-value of 0.001, while *Gender* is the least significant with a p-value of 0.069. Note that all the variables are significant at the 10% level, however.

(e) Four of the variables are significant at the 5% level: *ClassYear*, *CognitionZscore*, *DASScore* (just barely), and *Drinks*.

D.49 (a) This is an association between two quantitative variables, so can be tested with either a test for correlation or a test for slope in simple linear regression. Here we do a test for correlation. Let ρ be the true correlation between hours of exercise per week and GPA for all students, and our hypotheses are then

$$H_0 : \rho = 0$$
$$H_a : \rho \neq 0$$

Using technology, we find the sample correlation to be $r = -0.159$. We can use technology to find the p-value or we can use the formula. Using the formula, the t-statistic is then

$$t = \frac{r - 0}{\sqrt{\frac{1-r^2}{n-2}}} = \frac{-0.159}{\sqrt{\frac{1-(-0.159^2)}{343-2}}} = \frac{-0.159}{0.053} = -2.98.$$

We compare this to a t-distribution with df $= n - 2 = 343 - 2 = 341$, and get a p-value of 0.003. We have strong evidence for a negative correlation between hours of exercise per week and GPA for college students.

(b) Using technology, we fit the multiple regression model with *GPA* as the response variable and *Exercise* and *Gender* as explanatory variables. The output is below.

```
Coefficients:
             Estimate Std. Error t value Pr(>|t|)
(Intercept)  3.311599   0.043427  76.257  < 2e-16 ***
Exercise    -0.009739   0.003828  -2.544  0.01139 *
GenderCode  -0.125922   0.042497  -2.963  0.00326 **

Residual standard error: 0.3888 on 340 degrees of freedom
Multiple R-squared: 0.04967,    Adjusted R-squared: 0.04408
F-statistic: 8.885 on 2 and 340 DF,  p-value: 0.0001732
```

The p-value for testing the coefficient of *Exercise* of 0.011 indicates that *Exercise* is a significant predictor of *GPA*, even after accounting for *GenderCode*.

D.51 Here is some output for fitting the model to predict *Bodyfat* with all nine predictors.

The regression equation is Bodyfat = $-$ 23.7 + 0.0838 Age $-$ 0.0833 Weight + 0.036 Height + 0.001 Neck $-$ 0.139 Chest + 1.03 Abdomen + 0.226 Ankle + 0.148 Biceps $-$ 2.20 Wrist

```
Predictor     Coef  SE Coef      T      P
Constant    -23.66    29.46  -0.80  0.424
Age         0.08378  0.05066   1.65  0.102
Weight     -0.08332  0.08471  -0.98  0.328
Height       0.0359   0.2658   0.14  0.893
Neck         0.0011   0.3801   0.00  0.998
Chest       -0.1387   0.1609  -0.86  0.391
Abdomen      1.0327   0.1459   7.08  0.000
Ankle        0.2259   0.5417   0.42  0.678
Biceps       0.1483   0.2295   0.65  0.520
Wrist       -2.2034   0.8129  -2.71  0.008

S = 4.13552   R-Sq = 75.7%   R-Sq(adj) = 73.3%

Analysis of Variance
Source          DF       SS      MS      F      P
Regression       9  4807.36  534.15  31.23  0.000
Residual Error  90  1539.23   17.10
Total           99  6346.59
```

We see that a number of the predictors have very large p-values for the t-test of the coefficient, indicating that they are not very helpful in this model. Suppose that we drop the *Neck* and *Height* measurements which have the worst individual p-values (0.998 and 0.893) to obtain the new output shown below.

```
The regression equation is Bodyfat = - 20.5 + 0.0850 Age - 0.0757 Weight
    - 0.144 Chest + 1.02 Abdomen + 0.214 Ankle + 0.144 Biceps - 2.21 Wrist
```

Predictor	Coef	SE Coef	T	P
Constant	-20.47	14.62	-1.40	0.165
Age	0.08496	0.04938	1.72	0.089
Weight	-0.07569	0.05856	-1.29	0.199
Chest	-0.1444	0.1539	-0.94	0.350
Abdomen	1.0223	0.1231	8.30	0.000
Ankle	0.2137	0.5246	0.41	0.685
Biceps	0.1442	0.2244	0.64	0.522
Wrist	-2.2082	0.7485	-2.95	0.004

```
S = 4.09076   R-Sq = 75.7%   R-Sq(adj) = 73.9%
```

Analysis of Variance

Source	DF	SS	MS	F	P
Regression	7	4807.03	686.72	41.04	0.000
Residual Error	92	1539.56	16.73		
Total	99	6346.59			

We see that the value of $R^2 = 75.7\%$ remains unchanged and *SSModel* has only gone down by 0.33 (from 4807.36 to 4807.03). Clearly we lose essentially no predictive power by dropping these two very ineffective predictors, so the model is improved. But can you do even better?

When choosing a model from among many predictors there is rarely an absolute "best" choice that is optimal by all criteria. For this reason, there are several (even many) reasonable models that would be acceptable choices for this situation.

Final Essential Synthesis Solutions

E.1 *Dear Congressman Daniel Webster,*

You recently criticized the American Community Survey as being a random survey. However, the fact that it is a random survey is crucial for enabling us to make generalizations from the sample of people surveyed to the entire population of US residents. We can only generalize from the sample to the population if the sample is representative of the population (closely resembles the population in all characteristics, except only smaller). Unfortunately, without randomness we are notoriously bad at choosing representative samples. Because the whole point of the survey is to gain information about the population, we do not know what the population looks like, and so have no way of knowing what is "representative". On the bright side, randomly choosing a sample yields a group of people that are representative of the population. With a random sample, the larger the sample size, the closer the sample statistics will be to the population values you care about. With non-random samples this may not be the case. In short, we can best draw valid scientific conclusions from samples that have been randomly selected.

Sincerely,
A statistics student

E.3 (a) A bootstrap distribution, generated via StatKey and showing the cutoffs for the middle 90% is shown below:

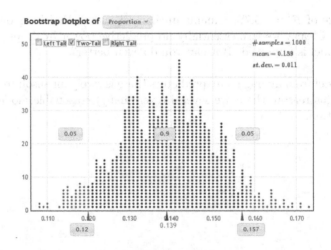

Based on the middle 90% of bootstrap proportions, a 90% confidence interval is 0.120 to 0.157.

(b) We estimate the standard error using the formula

$$SE = \sqrt{\frac{\hat{p}(1-\hat{p})}{n}} = \sqrt{\frac{0.139(1-0.139)}{1000}} = 0.011.$$

Notice that this matches the standard deviation of the bootstrap distribution. For a 90% confidence interval $z^* = 1.645$, so we generate the interval as

$$
\begin{array}{rcl}
\text{sample statistic} & \pm & z^* \cdot SE \\
\hat{p} & \pm & z^* \cdot SE \\
0.139 & \pm & 1.645 \cdot 0.011 \\
0.139 & \pm & 0.018 \\
0.121 & \text{to} & 0.157
\end{array}
$$

(c) We are 90% confident that between 12.1% and 15.7% of US residents do not have health insurance.

(d) The sample statistic is $\hat{p} = 0.139$ and the margin of error is $z^* \cdot SE = 1.645 \cdot 0.011 = 0.018$, or the distance from either bound of the confidence interval to the statistic, $0.157 - 0.139 = 0.018$.

(e) The sample size is much larger for the entire American Community Survey sample than for the 1000 people sub-sampled for **ACS**, so the margin of error will be much smaller.

(f) The 90% confidence interval is sample statistic $\pm$ margin of error, or $0.155 \pm 0.001 = (0.154, 0.156)$. Based on the full ACS survey, we are 90% confident that between 15.4% and 15.6% of all US residents do not have health insurance.

E.5   (a) *Income* is a quantitative variable, so we can visualize it's distribution with a histogram:

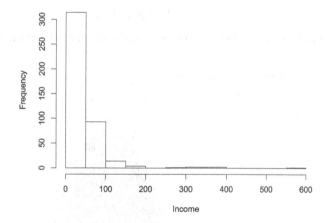

The distribution of income for employed US residents is strongly right-skewed. Most people have yearly incomes below \$100,000, but some people have incomes that are much higher. The maximum yearly income in this dataset is an outlier making about \$600,000 a year.

(b) The mean yearly income in the sample is $\bar{x} = \$41,494$, while the median income is \$29,000. The standard deviation of incomes is $s = \$52,248$ and the $IQR = \$40,900$. Yearly incomes in this dataset range from a minimum of \$0 to a maximum of \$563,000.

(c) We are looking at the relationship between a quantitative variable and a categorical variable, so can visualize with side-by-side boxplots:

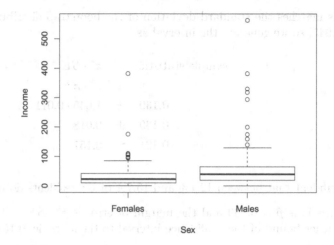

It appears that males tend to make more than females. The age distributions for each sex are heavily skewed towards larger incomes with several high outliers.

(d) In this sample, the males make an average of $\overline{x}_M = \$50,971$, while the females make an average of $\overline{x}_F = \$32,158$. The males make an average of $\$50,971 - \$32,158 = \$18,813$ more in income than the females.

(e) Let μ_M and μ_F denote the average yearly income for employed males and females, respectively, who live in the US. We test the hypotheses

$$H_0 : \mu_M = \mu_F$$
$$H_a : \mu_M \neq \mu_F.$$

We use StatKey or other technology to create a randomization distribution for difference in means, as shown below:

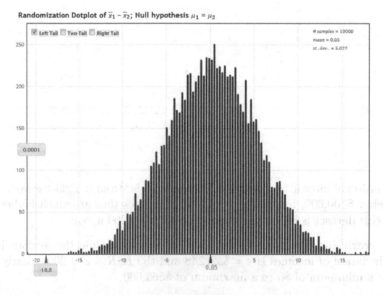

Of the 10,000 simulated randomization statistics, one one was less than the observed statistic (-18.8), so the area in one tail is 0.0001. We are testing a two-sided alternative, so double this to get a p-value of $2 \cdot 0.0001 = 0.0002$. This p-value is very small, so we have strong evidence that the average yearly income among employed US residents in 2010 is higher for males than for females.

E.7 (a) The number of people with health insurance in each racial group is the number of people of that race multiplied by the proportion of that race with health insurance. So the number of white people with health insurance is $761 \cdot 0.880 = 669.7$, which rounds to 670 (counts of people have to be whole numbers). The complete table is given below (and we could also generate this table directly from the **ACS** data).

	White	Black	Asian	Other
Health Insurance	670	86	59	46
No Health insurance	91	20	11	17

(b) This is a visualization of two categorical variables, which can be done with a segmented bar chart:

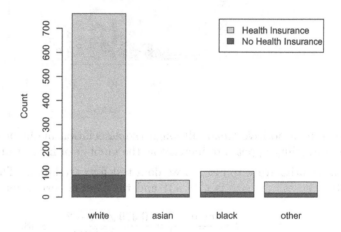

(c) We are testing for an association between two categorical variables, one of which has more than two levels, so use a chi-square test for association. The hypotheses are

H_0: There is no association between health insurance status and race
H_a: There is an association between health insurance status and race

We compute expected counts for each cell using (row total)(column total)/(sample size). The table of observed (expected) counts is given below:

	White	Black	Asian	Other	Total
Health Insurance	670 (655.2)	86 (91.27)	59 (60.3)	46 (54.2)	861
No Health insurance	91 (105.8)	20 (14.7)	11 (9.7)	17 (8.8)	139
Total	761	106	70	63	1000

This gives a chi-square statistic of

$$\chi^2 = \sum \frac{(observed - expected)^2}{expected} = \frac{(670 - 655.2)^2}{655.2} + \cdots + \frac{(17 - 8.8)^2}{8.8} = 13.79$$

The expected counts are all greater than 5, so we can compare this to a chi-square distribution with $(2-1)(4-1) = 2$ degrees of freedom, and find the p-value as the area above $\chi^2 = 13.79$. This gives a p-value of 0.0032. There is strong evidence for an association between whether or not a person has health insurance and race.

E.9 (a) *HoursWk* and *Income* are both quantitative variables, so we visualize with a scatterplot. Note: We could also switch the variables between the axes, unless we've read ahead to part (c).

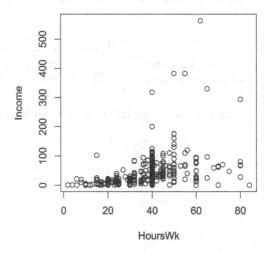

There appears to be a positive trend, although the association might be slightly curved, rather than linear, and the variability appears to increase as the number of hours worked increases.

(b) These are both quantitative variables, so we do a test for correlation. The hypotheses are $H_0 : \rho = 0$ vs $H_a : \rho > 0$. The sample size is $n = 431$ and the sample correlation is $r = 0.379$. The relevant t-statistic is

$$t = \frac{r\sqrt{n-2}}{\sqrt{1-r^2}} = \frac{0.379\sqrt{431-2}}{\sqrt{1-0.379^2}} = 8.48.$$

We compare this to a t-distribution with $431 - 2 = 429$ degrees of freedom, and find the p-value is essentially 0. Hours worked per week and income are very significantly positively associated.

(c) Some output for a regression model to predict *Income* based on *HoursWk* is given below:

```
Coefficients:
            Estimate Std. Error t value Pr(>|t|)
(Intercept) -18.3468     7.4349  -2.468    0.014 *
HoursWk       1.5529     0.1832   8.476 3.77e-16 ***

Residual standard error: 48.41 on 429 degrees of freedom
Multiple R-squared: 0.1435,    Adjusted R-squared: 0.1415
F-statistic: 71.85 on 1 and 429 DF,  p-value: 3.769e-16
```

The prediction equation is $\widehat{Income} = -18.3468 + 1.5529 \cdot HoursWk$.

(d) The predicted yearly income for someone who works 40 hours a week is

$$\widehat{Income} = -18.3468 + 1.5529 \cdot 40 = 43.769$$

or about \$43,769.

(e) The percent of the variability in income explained by the number of hours worked per week is $R^2 = 14.35\%$.

(f) Below is a scatterplot with the regression line on it:

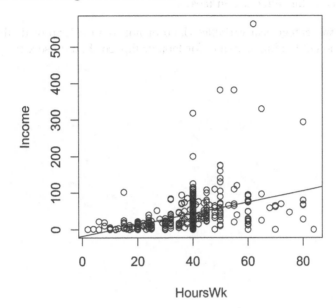

The condition of constant variability is clearly violated; variability in income is much higher when more hours are worked per week. There might also be a small amount of curvature in the relationship – note the relatively large number of points below the line for 20-35 hours of work a week and the negative predicted incomes when hours per week is very small.

E.11 Skull size is quantitative and which mound each skull was found in is categorical with two categories, so they should use a test for a difference in means.

E.13 We would like to make a statement about a population based on a sample proportion, so would use a confidence interval for a proportion.

E.15 We care about a single proportion (proportion of numbers ending in 0 or 5), and wish to determine whether this differs from 0.20, so would use a test for a proportion.

E.17 Both of these variables are quantitative, and the question is asking whether a negative correlation exists, so we would do a one-sided test for correlation. An lower-tail test for the slope in a regression model would also be acceptable.

E.19 This is testing for an association between two categorical variables, each with two categories, so we could use either a test for difference in proportions or a chi-square test for association.

E.21 We want to estimate a proportion, so would do an interval for a proportion.

E.23 We want to predict human equivalent age (quantitative) based on dog age (quantitative), so use slope of the simple linear regression.

E.25 Even though the experiment contains four groups, we are only interested in a categorical variable with two categories; exercise instructions or not, and one quantitative variable, amount of weight loss. The goal is not to test for a difference between the groups, but to estimate the difference between the groups, so we use a confidence interval for difference in means.

E.27 This involves two categorical variables (land or not, double loop or double flip), and she is interested whether the proportion of landing is higher for loop or flip, so should use a test for difference in proportions.

Section P.1 Solutions

P.1 We have $P(\text{not } A) = 1 - P(A) = 1 - 0.4 = 0.6$.

P.3 By the additive rule, we have $P(A \text{ or } B) = P(A) + P(B) - P(A \text{ and } B) = 0.4 + 0.3 - 0.1 = 0.6$.

P.5 We have
$$P(B \text{ if } A) = \frac{P(A \text{ and } B)}{P(A)} = \frac{0.1}{0.4} = 0.25.$$

P.7 We need to check whether $P(A \text{ and } B) = P(A) \cdot P(B)$. Since $P(A \text{ and } B) = 0.1$ and $P(A) \cdot P(B) = 0.4 \cdot 0.3 = 0.12$, the events are not independent. We can also check from an earlier exercise that $P(B \text{ if } A) = 0.25 \neq P(B) = 0.3$.

P.9 We have $P(\text{not } B) = 1 - P(B) = 1 - 0.4 = 0.6$.

P.11 We have
$$P(A \text{ if } B) = \frac{P(A \text{ and } B)}{P(B)} = \frac{0.25}{0.4} = 0.625.$$

P.13 No! If A and B were disjoint, that means they cannot both happen at once, which means $P(A \text{ and } B) = 0$. Since we are told that $P(A \text{ and } B) = 0.25$, not zero, the events are not disjoint.

P.15 Since A and B are independent, knowing that B occurs gives us no additional information about A, so $P(A \text{ if } B) = P(A) = 0.7$.

P.17 Since A and B are independent, we have $P(A \text{ and } B) = P(A) \cdot P(B) = 0.7 \cdot 0.6 = 0.42$.

P.19 There are two cells that are included as part of event A, so $P(A) = 0.2 + 0.1 = 0.3$.

P.21 There is one cell that is in both event A and event B, so we have $P(A \text{ and } B) = 0.2$.

P.23 We have
$$P(A \text{ if } B) = \frac{P(A \text{ and } B)}{P(B)} = \frac{0.2}{0.6} = 0.333.$$

P.25 No! If A and B were disjoint, that means they cannot both happen at once, which means $P(A \text{ and } B) = 0$. We see in the table that $P(A \text{ and } B) = 0.2$, not zero, so the events are not disjoint.

P.27 The two events are disjoint, since if at least one skittle is red then all three can't be green. However, they are not independent or complements.

P.29 The two events are independent, as Australia winning their rugby match will not change the probability that Poland wins their chess match. However, they are not disjoint or complements.

P.31   (a) It will not necessarily be the case that EXACTLY 1 in 10 adults are left-handed for every sample. We can only conclude that approximately 10% will be left-handed in the "long run" (for very large samples).

  (b) The three outcomes each have probability $\frac{1}{3}$ *only* if they are equally likely. This may not be the case for the results of baseball pitches.

(c) To find the probability of two consecutive 1's on independent dice rolls we should multiply the probabilities instead of adding them. Using the multiplicative rule, the probability that two consecutive rolls land with a 1 is $\frac{1}{6} \times \frac{1}{6} = \frac{1}{36}$.

(d) A probability that is not between 0 and 1 does not make sense.

P.33 (a) We are finding $P(MP)$. There are a total of 303 inductees and 206 of them are performers, so we have

$$P(MP) = \frac{206}{303} = 0.680$$

The probability that an inductee selected at random will be a performer is 0.680.

(b) We are finding $P(\text{not } F)$. There are a total of 303 inductees and 256 of them do not have any female members, so we have

$$P(\text{not } F) = \frac{256}{303} = 0.845$$

The probability that an inductee selected at random will not have any female members is 0.845.

(c) In this case, we are interested only in inductees who are performers, and we want to know the probability they have female members, $P(F \text{ if } MP)$. There are 206 performers and 38 of those have female members, so we have

$$P(F \text{ if } MP) = \frac{38}{206} = 0.184.$$

(d) In this case, we are interested only in inductees that do not have any female members, and we want to know the probability of not being a performer, $P(\text{not } MP \text{ if not } F)$. There are 256 inductees with no female members and 88 of them are not performers, so we have

$$P(\text{not } MP \text{ if not } F) = \frac{88}{256} = 0.344.$$

(e) We are finding $P(MP \text{ and not } F)$. Of the 303 inductees, there are 168 that are performers with no female members, so we have

$$P(MP \text{ and not } F) = \frac{168}{303} = 0.554$$

(f) We are finding $P(\text{not } MP \text{ or } F)$. Of the 303 inductees, $38 + 9 + 88 = 135$ are either not performers or have female members (or both), so we have

$$P(\text{not } MP \text{ or } F) = \frac{135}{303} = 0.446$$

Notice that this is the complement of the event found in part (e).

P.35 (a) There are 11 red ones out of a total of 80, so the probability that we pick a red one is $11/80 = 0.1375$.

(b) The probability that it *is* blue is $20/80 = 0.25$ so the probability that it is not blue is $1 - 0.25 = 0.75$.

(c) The single piece can be red or orange, but not both, so these are disjoint events. The probability the randomly selected candy is red or orange is $11/80 + 12/80 = 23/80 = 0.2875$.

(d) The probability that the first one is blue is $20/80 = 0.25$. When we put it back and mix them up, the probability that the next one is blue is also 0.25. By the multiplication rule, since the two selections are independent, the probability both selections are blue is $0.25 \cdot 0.25 = 0.0625$.

(e) The probability that the first one is red is $11/80$. Once that one is taken (since we don't put it back and we eat it instead), there are only 79 pieces left and 11 of those are green. By the multiplication rule, the probability of a red then a green is $(11/80) \cdot (11/79) = 0.191$.

P.37 Let S denote successfully making a free throw and F denote missing it.

(a) As free throws are independent, we can multiply the probabilities.
$$P(\text{Makes two}) = P(S_1 \text{ and } S_2) = P(S_1) \cdot P(S_2) = 0.908 \times 0.908 = 0.824$$

(b) The probability of missing one free throw is $P(F) = 1 - 0.908 = 0.092$. So,
$$P(\text{Misses two}) = P(F_1 \text{ and } F_2) = P(F_1) \cdot P(F_2) = 0.092 \times 0.092 = 0.008$$

(c) He can either miss the first and make the second shot, or make the first and miss the second. So,
$$P(\text{Makes exactly one}) = P(S_1 \text{ and } F_2) + P(F_1 \text{ and } S_2) = 0.908 \times 0.092 + 0.092 \times 0.908 = 0.167$$

P.39 (a) The probability that a women is not color-blind is $1 - 0.04 = 0.996$, and the probability that a man is not colorblind is $1 - 0.07 = 0.93$. As all events are independent, we can multiply their probabilities:
$$P(\text{Nobody is Colorblind}) = 0.996^{25} \times 0.93^{15} = 0.305$$

(b) The event "At least one is Colorblind" is the complement of the event "Nobody is Colorblind" which has its probability computed in part (a), so
$$P(\text{At least one is Colorblind}) = 1 - P(\text{Nobody is Colorblind}) = 1 - 0.305 = 0.695$$

(c) The probability that the randomly selected student is a man is $15/40 = 0.375$ and the probability that it is a women is $25/40 = 0.625$. Using the additive rule for disjoint events,
$$P(\text{Colorblind}) = P(\text{Colorblind Man OR Colorblind Woman})$$
$$= P(\text{Colorblind and Man}) + P(\text{Colorblind and Woman}).$$

By the multiplicative rule,
$$P(\text{Colorblind and Man}) = P(\text{Man}) \cdot P(\text{Colorblind if Man}) = 0.375 \times 0.07 = 0.02625$$

Similarly,
$$P(\text{Colorblind and Woman}) = 0.625 \times 0.004 = 0.0025$$

So, $P(\text{Colorblind}) = 0.02625 + 0.0025 = 0.029$.

P.41 (a) The probability that the S&P 500 increased on a randomly selected day is $423/756 = 0.5595$.

(b) Assuming independence, the probability that the S&P 500 increases for two consecutive days is $0.5595 \times 0.5595 = 0.3130$ (using the multiplicative rule). The probability that the S&P 500 increases on a day given that it increased the day before remains 0.5595 if the events are independent.

(c) The probability that the S&P 500 increases for two consecutive days is $234/755 = 0.3099$. The probability that the S&P 500 increases on a day, given that it increased on the previous day is
$$\frac{P(\text{Increase both days})}{P(\text{Increase first day})} = \frac{0.3099}{0.5595} = 0.5539$$

(d) The difference between the results in part (b) and part (c) is very small and insignificant, so we have little evidence that daily changes are not independent. (However, since the question does not ask for a formal hypothesis test, other answers are acceptable.)

Section P.2 Solutions

P.43 Here is the tree with the missing probabilities filled in.

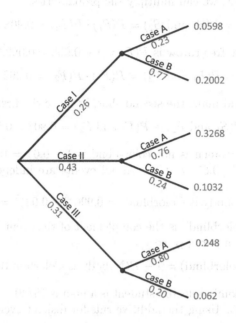

Note that the probabilities of all branches arising from a common point must sum to one, so

$$P(I) + P(II) + P(III) = 1 \implies P(I) = 1 - 0.43 - 0.31 = 0.26$$
$$P(A \text{ if } II) + P(B \text{ if } II) = 1 \implies P(A \text{ if } II) = 1 - 0.24 = 0.76$$

We obtain the probabilities at the end of each pair of branches with the multiplicative rule, so

$$P(II \text{ and } B) = P(II) \cdot P(B \text{ if } II) = 0.43(0.24) = 0.1032$$
$$P(III \text{ and } A) = P(III) \cdot P(A \text{ if } III) = 0.31(0.80) = 0.248$$

P.45 Here is the tree with the missing probabilities filled in.

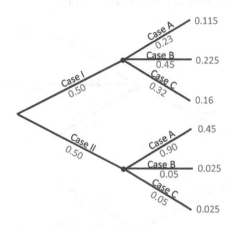

First, the sum of all the joint probabilities for all pairs of branches must be one so we have

$$P(A \text{ and } I) = 1 - (0.225 + 0.16 + 0.45 + 0.025 + 0.025) = 0.115$$

Using the total probability rule we have

$$P(I) \quad = \quad P(I \text{ and } A) + P(I \text{ and } B) + P(I \text{ and } C) = 0.115 + 0.225 + 0.16 = 0.5$$
$$P(II) \quad = \quad P(II \text{ and } A) + P(II \text{ and } B) + P(II \text{ and } C) = 0.45 + 0.025 + 0.025 = 0.5$$

The remaining six probabilities all come from the conditional probability rule. For example,

$$P(A \text{ if } I) = \frac{P(A \text{ and } I)}{P(I)} = \frac{0.115}{0.5} = 0.23$$

P.47 We use the multiplicative rule to see $P(\text{B and R}) = P(B) \cdot P(R \text{ if } B) = 0.4 \cdot 0.2 = 0.08$.

P.49 This conditional probability is shown directly on the tree diagram. Since we assume A is true, we follow the A branch and then find the probability of R, which we see is 0.9.

P.51 We see in the tree diagram that there are two ways for R to occur. We find these two probabilities and add them up, using the total probability rule. We see that $P(A \text{ and } R) = 0.6 \cdot 0.9 = 0.54$ so that top branch can be labeled 0.54. We also see that $P(B \text{ and } R) = 0.4 \cdot 0.2 = 0.08$. Since either A or B must occur (since these are the only two branches in this part of the tree), these are the only two ways that R can occur, so $P(R) = P(A \text{ and } R) + P(B \text{ and } R) = 0.54 + 0.08 = 0.62$.

P.53 We know that

$$P(\text{A if S}) = \frac{P(\text{A and S})}{P(\text{S})}$$

so we need to find $P(A \text{ and } S)$ and $P(S)$. Using the multiplicative rule, we see that $P(A \text{ and } S) = 0.6 \cdot 0.1 = 0.06$. We also see that $P(B \text{ and } S) = 0.4 \cdot 0.8 = 0.32$. By the total probability rule, we have

$$P(\text{A if S}) = \frac{P(\text{A and S})}{P(\text{S})} = \frac{P(\text{A and S})}{P(\text{A and S}) + P(\text{B and S})} = \frac{0.06}{0.06 + 0.32} = 0.158$$

We see that $P(\text{A if S}) = 0.158$.

P.55 We first create the tree diagram using the information given, and use the multiplication rule to fill in the probabilities at the ends of the branches. For example, for the top branch, the probability of having 1 occupant in an owner-occupied housing unit is $0.65 \cdot 0.217 = 0.141$.

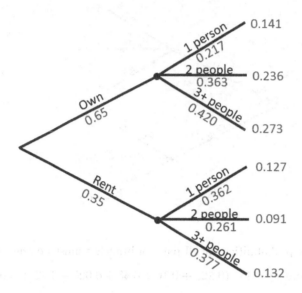

(a) We see at the end of the branch with rented and 2 occupants that the probability is 0.091.

(b) There are two branches that include having 3 or more occupants and we use the addition rule to see that the probability of 3 or more occupants is $0.273 + 0.132 = 0.405$.

(c) This is a conditional probability (or Bayes' rule). We have:

$$P(\text{rent if 1}) = \frac{P(\text{rent and 1 person})}{P(\text{1 person})} = \frac{0.127}{0.141 + 0.127} = \frac{0.127}{0.268} = 0.474$$

If a housing unit has only 1 occupant, the probability that it is rented is 0.474.

P.57 We are given

$$P(\text{Positive if no Cancer}) = 86.6/1000 = 0.0866,$$
$$P(\text{Positive if Cancer}) = 1 - 1.1/1000 = 0.9989, \text{ and}$$
$$P(\text{Cancer}) = 1/38 = 0.0263$$

Applying Bayes' rule we have

$$P(\text{Cancer if Positive}) = \frac{P(\text{Cancer})P(\text{Positive if Cancer})}{P(\text{no Cancer})P(\text{Positive if no Cancer}) + P(\text{Cancer})P(\text{Positive if Cancer})}$$
$$= \frac{0.0263 \cdot 0.9989}{(1 - 0.0263) \cdot 0.0866 + 0.0263 \cdot 0.9989}$$
$$= 0.2375$$

P.59 (a) Using the formula for conditional probability,

$$P(\text{Free if Spam}) = \frac{P(\text{Free and Spam})}{P(\text{Spam})} = \frac{0.0357}{0.134} = 0.266$$

(b) Using the formula for conditional probability,

$$P(\text{Spam if Free}) = \frac{P(\text{Free and Spam})}{P(\text{Free})} = \frac{0.0357}{0.0475} = 0.752$$

P.61 Using Bayes' rule,

$$P(\text{Spam if Text and Free}) = \frac{P(\text{Spam})P(\text{Text and Free if Spam})}{P(\text{Spam})P(\text{Text and Free if Spam}) + P(\text{not Spam})P(\text{Text and Free if not Spam})}$$
$$= \frac{0.134 \cdot 0.17}{0.134 \cdot 0.17 + 0.866 \cdot 0.0006}$$
$$= 0.978.$$

Section P.3 Solutions

P.63 A discrete random variable, as it can only take the values $\{0, 1, 2, \ldots, 10\}$.

P.65 A discrete random variable, as an ace must appear somewhere between the 1st and 48th card dealt.

P.67 A continuous random variable, as it can take any value greater than or equal to 0 lbs.

P.69 We see that $P(X = 3) = 0.2$ and $P(X = 4) = 0.1$ so $P(X = 3 \text{ or } X = 4) = 0.2 + 0.1 = 0.3$.

P.71 We have $P(X < 3) = P(X = 1 \text{ or } X = 2) = 0.4 + 0.3 = 0.7$.

P.73 We have $P(X \text{ is an even number}) = P(X = 2 \text{ or } X = 4) = 0.3 + 0.1 = 0.4$.

P.75 The probability values have to add up to 1.0 so we have $? = 1.0 - (0.2 + 0.2 + 0.2) = 0.4$.

P.77 The probability values have to add up to 1.0 but the sum of the values there is already greater than 1 (we see $0.3 + 0.3 + 0.3 + 0.3 = 1.2$). Since a probability cannot be a negative number, this cannot be a probability function.

P.79 (a) We multiply the values of the random variable (in this case, 10, 20, and 30) by the corresponding probability and add up the results. We have

$$\mu = 10(0.7) + 20(0.2) + 30(0.1) = 14$$

The mean of this random variable is 14.

(b) To find the standard deviation, we subtract the mean of 14 from each value, square the difference, multiply by the probability, and add up the results to find the variance; then take a square root to find the standard deviation.

$$\sigma^2 = (10 - 14)^2 \cdot 0.7 + (20 - 14)^2 \cdot 0.2 + (30 - 14)^2 \cdot 0.1 = 44$$
$$\implies \sigma = \sqrt{44} = 6.63$$

P.81 (a) We multiply the values of the random variable (in this case, 10, 12, 14, and 16) by the corresponding probability and add up the results. We have

$$\mu = 10(0.25) + 12(0.25) + 14(0.25) + 16(0.25) = 13$$

The mean of this random variable is 13, which makes sense since the probability distribution is symmetric and 13 is right in the middle.

(b) To find the standard deviation, we subtract the mean of 13 from each value, square the difference, multiply by the probability, and add up the results to find the variance; then take a square root to find the standard deviation.

$$\sigma^2 = (10 - 13)^2 \cdot 0.25 + (12 - 13)^2 \cdot 0.25 + (14 - 13)^2 \cdot 0.25 + (16 - 13)^2 \cdot 0.25 = 5$$
$$\implies \sigma = \sqrt{5} = 2.236$$

P.83 (a) We see that $0.362 + 0.261 + 0.153 + 0.114 + 0.061 + 0.027 + 0.022 = 1$, as expected.

(b) We have $p(1) + p(2) = 0.362 + 0.261 = 0.623$.

(c) We have $p(5) + p(6) + p(7) = 0.061 + 0.027 + 0.022 = 0.110$.

(d) It is easiest to find this probability using the complement rule, since more than 1 occupant is the complement of 1 occupant for this random variable. The answer is $1 - p(1) = 1 - 0.362 = 0.638$.

P.85 (a) We multiply the values of the random variable by the corresponding probability and add up the results. We have

$$\mu = 1(0.362) + 2(0.261) + 3(0.153) + 4(0.114) + 5(0.061) + 6(0.027) + 7(0.022) = 2.42$$

The average household size for a renter-occupied housing unit in the US is 2.42 people.

(b) To find the standard deviation, we subtract the mean of 2.42 from each value, square the difference, multiply by the probability, and add up the results to find the variance; then take a square root to find the standard deviation.

$$\begin{aligned}\sigma^2 &= (1 - 2.42)^2 \cdot 0.362 + (2 - 2.42)^2 \cdot 0.261 + \cdots + (7 - 2.42)^2 \cdot 0.022 \\ &= 2.3256 \\ \Longrightarrow \sigma &= \sqrt{2.3256} = 1.525\end{aligned}$$

P.87 (a) As the probabilities must sum to 1, so

$$P(X = 4) = 1 - (0.29 + 0.3 + 0.2 + 0.17) = 1 - 0.96 = 0.04$$

(b) $P(X < 2) = P(X = 0) + P(X = 1) = 0.29 + 0.3 = 0.59$

(c) To find the mean we use

$$\mu = 0 \cdot 0.29 + 1 \cdot 0.3 + 2 \cdot 0.2 + 3 \cdot 0.17 + 4 \cdot 0.04 = 1.37 \text{ cars}$$

(d) To find the standard deviation we use

$$\begin{aligned}\sigma^2 &= (0 - 1.37)^2 0.29 + (1 - 1.37)^2 0.3 + (2 - 1.37)^2 0.2 + (3 - 1.37)^2 \cdot 0.17 + (4 - 1.37)^2 0.04 \\ &= 1.393 \\ \Longrightarrow \sigma &= \sqrt{1.393} = 1.180 \text{ cars}\end{aligned}$$

P.89 Let X_1, X_2, and X_3 represent the sales on each of three consecutive days. Since daily sales are independent, we can multiple their probabilities. So, the probability that the no cars are sold in three consecutive days is

$$\begin{aligned}P(X_1 = 0 \text{ and } X_2 = 0 \text{ and } X_3 = 0) &= P(X_1 = 0) \cdot P(X_2 = 0) \cdot P(X_3 = 0) \\ &= 0.29 \cdot 0.29 \cdot 0.29 \\ &= 0.29^3 \\ &= 0.0244\end{aligned}$$

P.91 (a) If the woman dies during the first year the organization loses $\$100000 - c$, if the woman dies during the second year the organization loses $\$100000 - 2c$, and so on. If the woman does not die during the five year contract the organization earns $5c$ dollars, and the probability of this is $1 - (0.00648 +$

0.00700 + 0.00760 + 0.00829 + 0.00908) = 0.96155. The probability distribution of the profit (as a function of the yearly fee c) is given below.

x	$c - \$100000$	$2c - \$100000$	$3c - \$100000$	$4c - \$100000$	$5c - \$100000$	$5c$
$p(x)$	0.00648	0.00700	0.00760	0.00829	0.00908	0.96155

(b) In terms of c,

$$\mu = (c - \$100000) \cdot 0.00648 + (2c - \$100000) \cdot 0.00700 + \ldots + (5c - \$100000) \cdot 0.00908 + 5c \cdot 0.96155$$
$$= 4.9296c - \$3845$$

(c) Setting $\mu = 0$ and solving, we get $c = \frac{3845}{4.9296} = 779.98$, so the organization would have to charge approximately \$779.98 per year.

P.93   (a) We find the probability an street address starts with "1" as

$$P(X = 1) = log_{10}(1 + 1/1) = log_{10}(2) = 0.301$$

For an address starting with the digit "9" we have

$$P(X = 1) = log_{10}(1 + 1/9) = log_{10}(1.111\ldots) = 0.046$$

(b) To find $P(X > 2)$ we use the complement rule to find

$$\begin{aligned} P(X > 2) &= 1 - [P(X = 1) + P(X = 2)] \\ &= 1 - [log_{10}(1 + 1/1) + log10(1 + 1/2)] \\ &= 1 - [log_{10}(2) + log_{10}(1.5)] \\ &= 1 - [0.301 + 0.176] \\ &= 1 - 0.477 = 0.523 \end{aligned}$$

Section P.4 Solutions

P.95 This is a binomial random variable, with $n = 10$ and $p = 1/6$.

P.97 Not binomial, since it is not clear what is counted as a success.

P.99 This is a binomial random variable, with $n = 100$ and $p = 0.51$.

P.101 We see that $7! = 7 \cdot 6 \cdot 5 \cdot 4 \cdot 3 \cdot 2 \cdot 1 = 5040$.

P.103 We see that $6! = 6 \cdot 5 \cdot 4 \cdot 3 \cdot 2 \cdot 1 = 720$.

P.105 We have $\binom{5}{2} = \dfrac{5!}{2!(3!)} = \dfrac{5 \cdot 4 \cdot 3 \cdot 2 \cdot 1}{2 \cdot 1 \cdot 3 \cdot 2 \cdot 1} = \dfrac{20}{2} = 10$.

P.107 We have $\binom{6}{5} = \dfrac{6!}{5!(1!)} = \dfrac{6 \cdot 5 \cdot 4 \cdot 3 \cdot 2 \cdot 1}{5 \cdot 4 \cdot 3 \cdot 2 \cdot 1 \cdot 1} = \dfrac{6}{1} = 6$.

P.109 We first calculate that $\binom{8}{7} = \dfrac{8!}{7!(1!)} = 8$. We then find

$$P(X = 7) = \binom{8}{7}(0.9^7)(0.1^1) = 8(0.9^7)(0.1^1) = 0.383.$$

P.111 We first calculate that $\binom{12}{8} = \dfrac{12!}{8!(4!)} = 495$. We then find

$$P(X = 8) = \binom{12}{8}(0.75^8)(0.25^4) = 495(0.75^8)(0.25^4) = 0.194.$$

P.113 The mean is $\mu = np = 10(0.8) = 8$ and the standard deviation is

$$\sigma = \sqrt{np(1 - p)} = \sqrt{10(0.8)(0.2)} = \sqrt{1.6} = 1.265$$

P.115 The mean is $\mu = np = 800(0.25) = 200$ and the standard deviation is

$$\sigma = \sqrt{np(1 - p)} = \sqrt{800(0.25)(0.75)} = \sqrt{150} = 12.25$$

P.117 A probability function gives the probability for each possible value of the random variable. There are five possible values for this random variable: 0 seniors, 1 senior, 2 seniors, 3 seniors, or all 4 seniors. This is a binomial random variable with $n = 4$ and $p = 0.25$.
The probability that none of the students are seniors is:

$$P(X = 0) = \binom{4}{0}(0.25^0)(0.75^4) = 1 \cdot 1 \cdot 0.75^4 = 0.316$$

The probability of 1 senior is:

$$P(X = 1) = \binom{4}{1}(0.25^1)(0.75^3) = 4 \cdot (0.25^1)(0.75^3) = 0.422$$

The probability of 2 seniors is:

$$P(X = 2) = \binom{4}{2}(0.25^2)(0.75^2) = 6 \cdot (0.25^2)(0.75^2) = 0.211$$

The probability of 3 seniors is:

$$P(X = 3) = \binom{4}{3}(0.25^3)(0.75^1) = 4 \cdot (0.25^3)(0.75^1) = 0.047$$

The probability that all 4 students are seniors is:

$$P(X = 4) = \binom{4}{4}(0.25^4)(0.75^0) = 1 \cdot (0.25^4) \cdot 1 = 0.004$$

We can summarize these results with a table for the probability function.

x	0	1	2	3	4
$p(x)$	0.316	0.422	0.211	0.047	0.004

Notice that the five probabilities add up to 1, as we expect for a probability function.

P.119 If X is the random variable giving the number of senior citizens (aged 65 or older) in a random sample of 10 people in the US, then X is a binomial random variable with $n = 10$ and $p = 0.13$. We are finding $P(X = 3)$ and $P(X = 4)$. To find $P(X = 3)$, we first calculate

$$\binom{10}{3} = \frac{10!}{3!(7!)} = 120$$

We then find

$$P(X = 3) = \binom{10}{3}(0.13^3)(0.87^7) = 120(0.13^3)(0.87^7) = 0.099.$$

To find $P(X = 4)$, we first calculate

$$\binom{10}{4} = \frac{10!}{4!(6!)} = 210$$

We then find

$$P(X = 4) = \binom{10}{4}(0.13^4)(0.87^6) = 210(0.13^4)(0.87^6) = 0.026.$$

The probability is 0.099 that 3 of the 10 people are senior citizens and is 0.026 that 4 of them are senior citizens.

P.121 The mean is $\mu = np = 3(0.49) = 1.47$ girls and the standard deviation is

$$\sigma = \sqrt{np(1-p)} = \sqrt{3(0.49)(0.51)}\sqrt{.7497} = 0.866$$

P.123 This is a binomial random variable with $n = 12$ and $p = 0.275$.

The mean is $\mu = np = 12(0.275) = 3.3$ college graduates in a sample of 12.

The standard deviation is $\sigma = \sqrt{np(1-p)} = \sqrt{12(0.275)(0.725)} = \sqrt{2.3925} = 1.55$.

P.125 This is a binomial random variable with $n = 20$ and $p = 0.65$.

The mean is $\mu = np = 20(0.65) = 13.0$ owner-occupied units in a sample of 20.

The standard deviation is $\sigma = \sqrt{np(1-p)} = \sqrt{20(0.65)(0.35)} = \sqrt{4.55} = 2.13$.

P.127 Let X measure the number of passengers (out of 32) who show up for a flight. For each passenger we have a 90% chance of showing up, so X is a binomial random variable with $n = 32$ and $p = 0.90$.

(a) The mean number of passengers on each flight is $\mu = np = 32(0.9) = 28.8$ people.

(b) Everyone gets a seat when $X \leq 30$. To find this probability we use the complement rule (find the chance too many people show up with $X = 31$ or $X = 32$, then subtract from one.)

$$
\begin{aligned}
P(X \leq 30) &= 1 - [P(X = 31) + P(X = 32)] \\
&= 1 - \left[\binom{32}{31} 0.9^{31} 0.1^1 + \binom{32}{0} 0.9^{32} 0.1^0 \right] \\
&= 1 - [32 \cdot 0.9^{31}(0.1) + 1 \cdot 0.9^{32} \cdot 1] \\
&= 1 - [0.122 + 0.034] \\
&= 1 - 0.156 \\
&= 0.844
\end{aligned}
$$

Everyone will have a seat on about 84.4% of the flights. The airline will need to deal with overbooked passengers on the other 15.6% of the flights.

Section P.5 Solutions

P.129 The area for values below 25 is clearly more than half the total area, but not as much as 95%, so 62% is the best estimate.

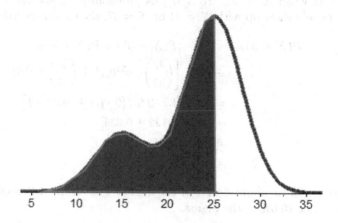

P.131 Almost all of the area under the density curve is between 10 and 30, so 95% is the best estimate.

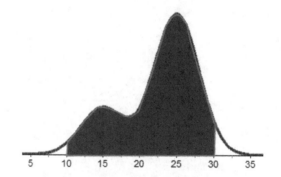

P.133 The plots below show the three required regions as areas in a $N(0, 1)$ distribution. We see that the areas are

(a) 0.8508

(b) 0.9332

(c) 0.1359

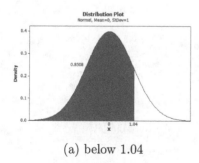

(a) below 1.04

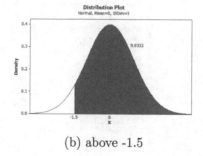

(b) above -1.5

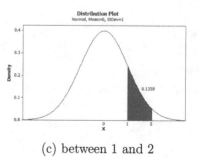

(c) between 1 and 2

P.135 The plots below show the three required regions as areas in a N(0,1) distribution. We see that the areas are

(a) 0.982

(b) 0.309

(c) 0.625

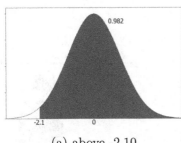

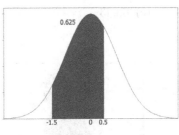

(a) above -2.10   (b) below -0.5   (c) between -1.5 and 0.5

P.137 The plots below show the required endpoint(s) for a $N(0,1)$ distribution. We see that the endpoint z is

(a) -1.282

(b) -0.8416

(c) ± 1.960

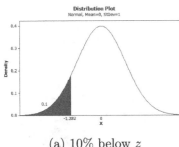

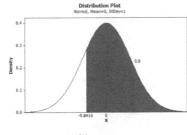

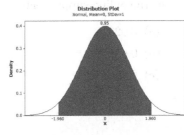

(a) 10% below z 80% above z 95% between $-z$ and $+z$

P.139 The plots below show the required endpoint(s) for a N(0,1) distribution. We see that the endpoint z is

(a) -1.28

(b) 0.385

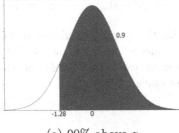

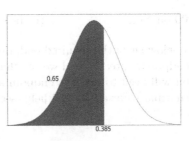

(a) 90% above z (b) 65% below z

P.141 The plots below show the three required regions as areas on the appropriate normal curve. Using technology, we can find the areas directly, and we see that the areas are as follows. (If you are using a paper table, you will need to first convert the values to the standard normal using z-scores.) For additional help, see the online supplements.

(a) 0.691

(b) 0.202

(c) 0.643

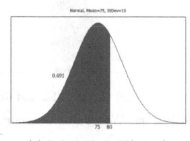 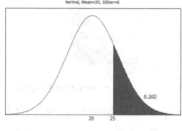 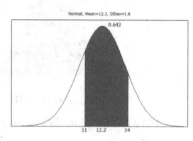

(a) below 80 on N(75,10)   (b) above 25 on N(20,6)   (c) between 11 and 14 on N(12.2, 1.6)

P.143 The plots below show the three required regions as areas on the appropriate normal curve. Using technology, we can find the areas directly, and we see that the areas are as follows. (If you are using a paper table, you will need to first convert the values to the standard normal using z-scores.) For additional help, see the online supplements.

(a) 0.023

(b) 0.006

(c) 0.700

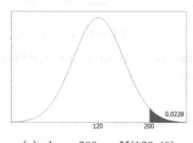 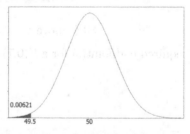 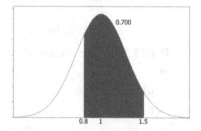

(a) above 200 on N(120,40)   (b) below 49.5 on N(50,0.2)   (c) between 0.8 and 1.5 on N(1,0.3)

P.145 The plots below show the required endpoint(s) for the given normal distribution. Using technology, we can find the endpoints directly, and we see that the requested endpoints are as follows. (If you are using a paper table, you will need to find the endpoints on a standard normal table and then convert them back to the requested normal.) For additional help, see the online supplements.

(a) 59.3

(b) 2.03

(c) 60.8 and 139.2. Notice that this is very close to our rough rule that about 95% of a normal distribution is within 2 standard deviations of the mean.

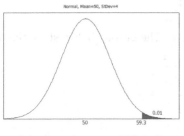
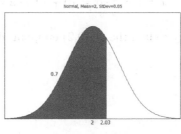
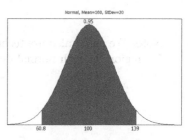

(a) 0.01 above on N(50,4)   (b) 0.70 below on N(2,0.05)   (c) 95% between on N(100,20)

P.147 The plots below show the required endpoint(s) for the given normal distribution. Using technology, we can find the endpoints directly, and we see that the requested endpoints are as follows. (If you are using a paper table, you will need to find the endpoints on a standard normal table and then convert them back to the requested normal.) For additional help, see the online supplements.

(a) 110

(b) 9.88

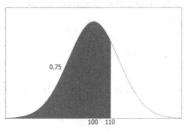
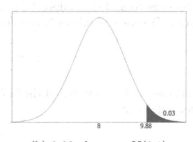

(a) 0.75 below on N(100,15)   (b) 0.03 above on N(8,1)

P.149 We standardize the endpoint of 40 using a mean of 48 and standard deviation of 5 to get

$$z = \frac{x - \mu}{\sigma} = \frac{40 - 48}{5} = -1.6$$

The graphs below show the lower tail region on each normal density. The shaded area in both curves is 0.0548.

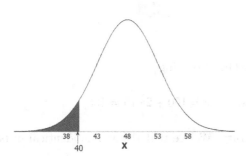
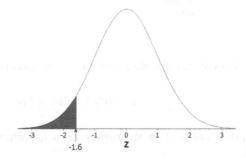

P.151 We use technology to find the endpoint on a standard normal curve that has 5% above it. The graph below shows this point is $z = 1.64$. We then convert this to a N(10,2) endpoint with

$$x = \mu + z \cdot \sigma = 10 + 1.64 \cdot 2 = 13.3$$

Note: We could also use technology to find the N(10,2) endpoint directly. The graphs below show the upper 5% region on each normal density.

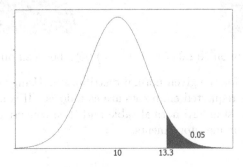 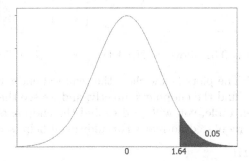

P.153 Using technology to find the endpoint for a standard normal density, we see that it is $z = -1.28$ (as shown in the figure). We convert this to N(500,80) using

$$x = \mu + z \cdot \sigma = 500 - 1.28 \cdot 80 = 397.6$$

Note: We could also use technology to find the N(500,80) endpoint directly. The graphs show the lower 10% region on each normal density.

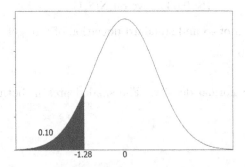 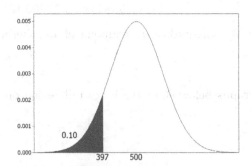

P.155 We convert the standard normal endpoints to N(100,15) using

$$x = 100 + 1 \cdot 15 = 115 \qquad \text{and} \qquad x = 100 + 2 \cdot 15 = 130$$

The graphs below show the region on each normal density. We see that the area is identical in both and is 0.1359.

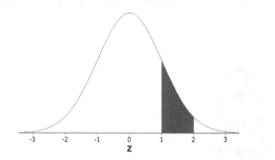 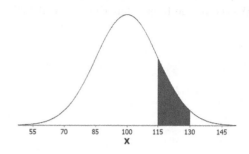

P.157 For Critical Reading, we have a $N(495, 116)$ curve, which will be centered at its mean, 495. We label the points that are one standard deviation away from the mean (379 and 611) and two standard deviations ways (263 and 727) so that approximately 95% of the area falls between the values two standard deviations away. Those points should be out in the tails, with only about 2.5% of the distribution beyond them on each side. See the figure.

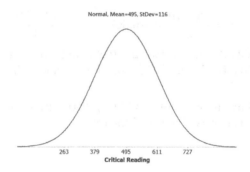

P.159 The plots below show the required endpoint(s) and/or probabilities for the given normal distributions. Note that a percentile always means the area to the left. Using technology, we can find the endpoints and areas directly, and we obtain the answers below. (Alternately, we could convert to a standard normal and use the standard normal to find the equivalent area.)

(a) The area below 450 is 0.384, so that point is about the 38th percentile of a N(484,115) distribution.

(b) The point where 90% of the scores are below it is a score of 631.

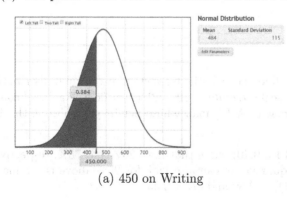 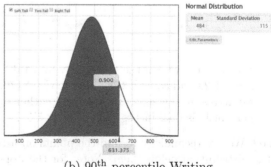

(a) 450 on Writing   (b) 90th percentile Writing

P.161 (a) Using technology we find the area between 68 inches and 72 inches in a $N(70,3)$ distribution. We see in the figure that the area is 0.495.

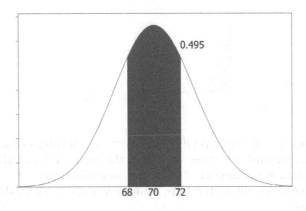

Alternately, we can compute z-scores:

$$z = \frac{68-70}{3} = -0.667 \qquad z = \frac{72-70}{3} = 0.667$$

Using technology or a table the area between -0.667 and 0.667 on a standard normal curve is 0.495, matching what we see directly. About 49.5% (or almost exactly half) of US men are between 68 and 72 inches tall.

(b) Using technology we find an endpoint for a $N(70,3)$ distribution that has an area of 0.10 below it. We see in the figure that the height is about 66.2 inches.

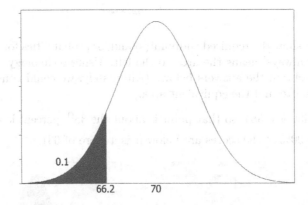

Alternately, we can use technology or a table to find the 10%-tile for a standard normal distribution, $z = -1.28$ and convert this value to the $N(70,3)$ scale, $Height = 70 - 1.28(3) = 66.2$. Again, of course, we arrive at the same answer using either approach. A US man whose height puts him at the 10th percentile is 66.2 inches tall or about 5'6".

P.163 We use technology to find the points on a N(3.16, 0.40) curve that have 25% and 75%, respectively, of the distribution below them. Note: for the upper quartile we can also look for 25% above the value. In the figure below we see that the two quartiles are at $Q_1 = 2.89$ and $Q_3 = 3.43$.

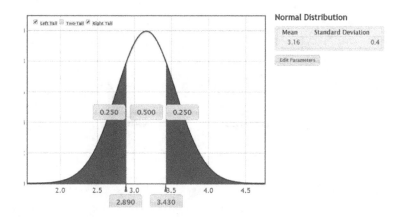

Alternately, we can use technology or a table to find the 25% and 75% endpoints for a standard normal distribution, $z = \pm 0.674$. We then convert these endpoints to corresponding points on a $N(3.16, 0.40)$ distribution.

$$Q_1 = 3.16 - 0.674 \cdot 0.40 = 2.89$$
$$Q_3 = 3.16 + 0.674 \cdot 0.40 = 3.43$$

P.165 The plots below show the three required regions as areas in a N(21.97,0.65) distribution. We see that the areas are 0.0565, 0.0012, and 0.5578, respectively.

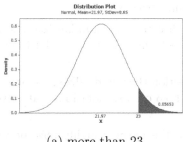
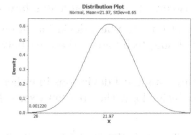
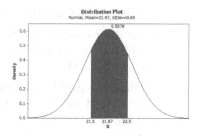

(a) more than 23　　　　　(b) less than 20　　　　　(c) between 21.5 and 22.5

If converting to a standard normal, the relevant z-scores and areas are shown below.

(a) $z = \frac{23 - 21.97}{0.65} = 1.585$. The area above 1.585 for $N(0,1)$ is 0.0565.

(b) $z = \frac{20 - 21.97}{0.65} = -3.031$. The area below -3.031 for $N(0,1)$ is 0.0012.

(c) $z = \frac{21.5 - 21.97}{0.65} = -0.7231$ and $z = \frac{22.5 - 21.97}{0.65} = 0.8154$. The area between -0.7231 and 0.8154 for $N(0,1)$ is 0.5578.

P.167 We use technology to determine the answers. We see in the figure that the results are:

(a) 0.0509 or 5.09% of students scored above a 90.

(b) 0.138 or 13.8% of students scored below a 60.

(c) Students with grades below 53.9 will be required to attend the extra sessions.

(d) Students with grades above 86.1 will receive a grade of A.

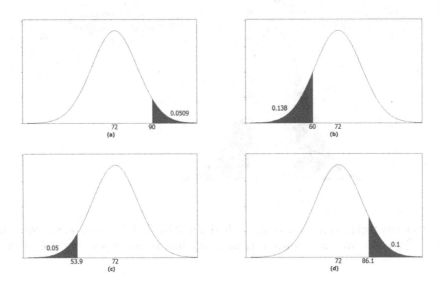

P.169 (a) For any $N(\mu, \sigma)$ distribution, when we standardize $\mu - 2\sigma$ and $\mu + 2\sigma$ we must get $z = -2$ and $z = +2$. For example, if we use $N(100, 20)$ the interval within two standard deviations of the mean goes from 60 to 140.

$$z = \frac{60 - 100}{20} = -2 \qquad \text{and} \qquad z = \frac{140 - 100}{20} = +2$$

Using technology, the area between -2 and $+2$ on a standard normal curve is 0.954.

(b) Similar to part (a), if we go just one standard deviation in either direction the standardized z-scores will be $z = -1$ and $z = +1$. Using technology, the area between -1 and $+1$ on a standard normal curve is 0.683.

(c) Similar to part (a), if we go three standard deviations in either direction the standardized z-scores will be $z = -3$ and $z = +3$. Using technology, the area between -3 and $+3$ on a standard normal curve is 0.997.

(d) The percentages within one, two, or three standard deviations of the mean, roughly 68% between $\mu \pm \sigma$, roughly 95% between $\mu \pm 2\sigma$, and roughly 99.7% between $\mu \pm 3\sigma$, should hold for any normal distribution since the standardized z-scores will always be $z = \pm 1$ or $z = \pm 2$ or $z = \pm 3$, respectively.